**Michael S. Hildebrand, Gregory G. Noll,** *and* **William Hand**

# Intermodal Container Emergencies

**SECOND EDITION**

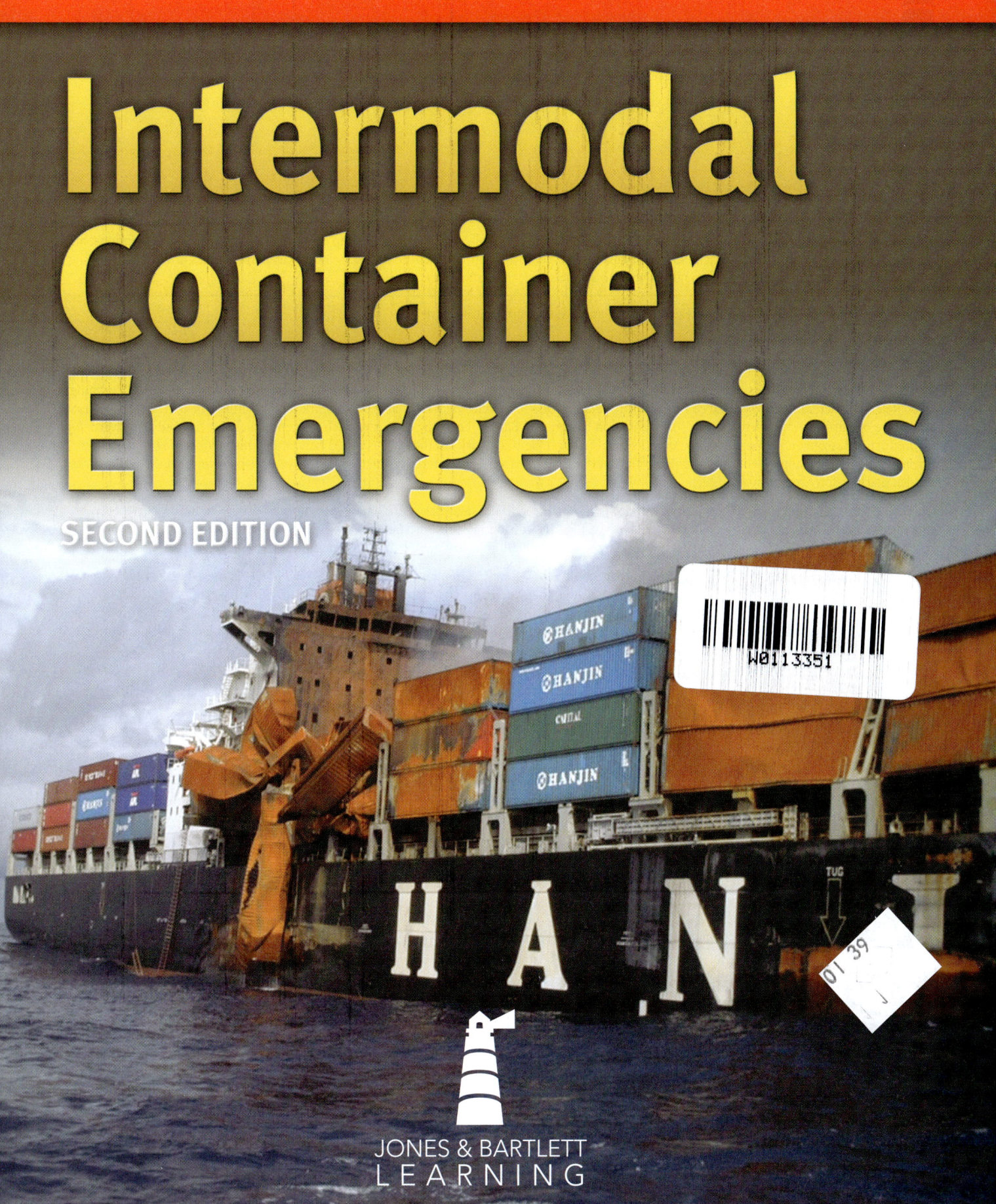

JONES & BARTLETT
LEARNING

*World Headquarters*
Jones & Bartlett Learning
5 Wall Street
Burlington, MA 01803
978-443-5000
info@jblearning.com
www.jblearning.com

Jones & Bartlett Learning books and products are available through most bookstores and online booksellers. To contact Jones & Bartlett Learning directly, call 800-832-0034, fax 978-443-8000, or visit our website, www.jblearning.com.

Substantial discounts on bulk quantities of Jones & Bartlett Learning publications are available to corporations, professional associations, and other qualified organizations. For details and specific discount information, contact the special sales department at Jones & Bartlett Learning via the above contact information or send an email to specialsales@jblearning.com.

## Production Credits

General Manager, Safety and Trades: Doug Kaplan
General Manager, Executive Publisher: Kimberly Brophy
VP, Product Development and Executive Editor: Christine Emerton
Director, PSG Editorial Development: Carol B. Guerrero
Executive Editor: Bill Larkin
Business Development Manager: Janet Maker
Editorial Assistant: Jessica Sturtevant
Vendor Manager: Nora Menzi
VP, Sales, Public Safety Group: Matthew Maniscalco
Director of Sales, Public Safety Group: Patricia Einstein
Director of Ecommerce and Digital Marketing: Eric Steeves

Director of Marketing Operations: Brian Rooney
VP, Manufacturing and Inventory Control: Therese Connell
Composition and Project Management: Integra Software Services Pvt. Ltd.
Cover Design: Kristin E. Parker
Director of Rights & Media: Joanna Gallant
Rights & Media Specialist: Robert Boder
Media Development Editor: Shannon Sheehan
Cover Image: © Alert Disaster Control (ALERT)
Printing and Binding: LSC Communications
Cover Printing: LSC Communications

**Library of Congress Cataloging-in-Publication Data unavailable at time of printing.**

6048
Printed in the United States of America
21 20 19 18 17    10 9 8 7 6 5 4 3 2 1

## Notice

Hazardous materials emergency response work is extremely dangerous; responders have died or sustained serious injury and illness while attempting to mitigate an incident. There is no possible way that this text can cover the full spectrum of problems and contingencies for dealing with every type of emergency incident. The user is warned to exercise all necessary cautions when engaged in hazardous materials emergency response. Always assume a worst-case scenario and place personal safety first.

It is the intent of the authors that this text be a part of the user's formal training in the response to hazardous materials emergencies involving intermodal containers. Even though this book is based on commonly used practices, references, laws, regulations, and consensus standards, it is not meant to set a standard of operations for any emergency response organization. The users are directed to develop their own Standard Operating Procedures and follow all system, agency, or employer guidelines for handling hazardous materials. It is the user's sole responsibility to stay up to date with procedures, regulations, and product developments that may improve personal health and safety.

No amount of technical knowledge assembled at the scene of an emergency can guarantee a safe and effective response. As is the case with any hazardous materials emergency, the best way to assure the safety of personnel working at the scene of an incident is to handle the emergency using the Incident Command System while following Standard Operating Procedures under the oversight of the incident commander and a safety officer.

# Table of Contents

© Photos.com/Getty.

Acknowledgments . . . . . . . . . . . . . . . . . . . . . . . . . . . . vii
About the Authors. . . . . . . . . . . . . . . . . . . . . . . . . . . viii

**CHAPTER 1**  Introduction. . . . . . . . . . . . . . . . . . . . 1

Chapter Outline . . . . . . . . . . . . . . . . . . . . . .1
Key Terms . . . . . . . . . . . . . . . . . . . . . . . . .1
Introduction . . . . . . . . . . . . . . . . . . . . . . . .1
Target Audience . . . . . . . . . . . . . . . . . . . . . .2
Overview of the Intermodal Container
    Industry . . . . . . . . . . . . . . . . . . . . . . .2
    Purpose of an Intermodal
        Container . . . . . . . . . . . . . . . . . .2
    The Land Bridge Concept . . . . . . . .4
Objectives . . . . . . . . . . . . . . . . . . . . . . . .4
Summary. . . . . . . . . . . . . . . . . . . . . . . . . .7
References . . . . . . . . . . . . . . . . . . . . . . . .7

**CHAPTER 2**  Codes, Standards, and
Regulations . . . . . . . . . . . . . . . . . . . . 8

Chapter Outline . . . . . . . . . . . . . . . . . . . . . .8
Key Terms . . . . . . . . . . . . . . . . . . . . . . . . .8
Introduction . . . . . . . . . . . . . . . . . . . . . . . .8
United Nations Recommendations
    on the Transport of Dangerous
    Goods—Model Regulations . . . . . . . .8
International Maritime Dangerous
    Goods (IMDG) Code. . . . . . . . . . . . . .9
U.S. Department of Transportation
    (U.S. DOT). . . . . . . . . . . . . . . . . . . .10
Transport Canada (TC) . . . . . . . . . . . . . .10
Normas Oficiales Mexicanas (NOM) . . . .10
Association of American Railroads
    (AAR). . . . . . . . . . . . . . . . . . . . . . .10
International Standards Organization
    (ISO) . . . . . . . . . . . . . . . . . . . . . . .11
Convention for Safe Containers (CSC) . . 11
Transportation Security Regulations . . . 11

Summary. . . . . . . . . . . . . . . . . . . . . . . . .12
References . . . . . . . . . . . . . . . . . . . . . . .12

**CHAPTER 3**  Intermodal Container Marking
Systems and Hazardous Materials
Placards. . . . . . . . . . . . . . . . . . . . . . 13

Chapter Outline . . . . . . . . . . . . . . . . . . . .13
Key Terms . . . . . . . . . . . . . . . . . . . . . . . .13
Introduction . . . . . . . . . . . . . . . . . . . . . .13
Container Markings and Identification
    Data Plates. . . . . . . . . . . . . . . . . . .13
Placards. . . . . . . . . . . . . . . . . . . . . . . . .16
Other Types of Intermodal container
    Identification Systems . . . . . . . . . . .16
    ADR/RID Hazard Marking
        System . . . . . . . . . . . . . . . . . .16
    HAZCHEM Marking System . . . . . .18
Markings Unique to Portable Tank
    Containers . . . . . . . . . . . . . . . . . . .18
    AAR 600 Marking . . . . . . . . . . . . .19
    Optional Markings . . . . . . . . . . . .22
Summary. . . . . . . . . . . . . . . . . . . . . . . . .29
References . . . . . . . . . . . . . . . . . . . . . . .29

**CHAPTER 4**  Intermodal Freight Containers . . . .30

Chapter Outline . . . . . . . . . . . . . . . . . . . .30
Key Terms . . . . . . . . . . . . . . . . . . . . . . . .30
Introduction . . . . . . . . . . . . . . . . . . . . . .30
Freight Container Construction
    Features . . . . . . . . . . . . . . . . . . . . .30
    Materials of Construction . . . . . . .31
    Corner Castings. . . . . . . . . . . . . . .33
    Container Floors . . . . . . . . . . . . . .34
    Cargo Capacities and Weights . . . .34
    Freight Container Cargo
        Packaging Systems . . . . . . . . . .34

Nonbulk Packaging . . . . . . . . . . . . . . . . .35
    Bags. . . . . . . . . . . . . . . . . . . . . . .35
    Drums . . . . . . . . . . . . . . . . . . . . .36
    Boxes. . . . . . . . . . . . . . . . . . . . .37
Bulk Packaging That May Be Placed
    on/in Transport Vehicles. . . . . . . . . . .37
    Flexible Intermediate Bulk
        Container (FIBC). . . . . . . . . . . .37
    Rigid Intermediate Bulk
        Containers (IBCs). . . . . . . . . . .38
Summary. . . . . . . . . . . . . . . . . . . . . . .42
References . . . . . . . . . . . . . . . . . . . . . .42

**CHAPTER 5 Nonpressure Intermodal Tank Containers . . . . . . . . . . . . . . . . . . . . . 43**

Chapter Outline . . . . . . . . . . . . . . . . . .43
Key Terms . . . . . . . . . . . . . . . . . . . . . .43
Introduction to Intermodal Tank
    Containers In General . . . . . . . . . . . .43
Design and Construction Features . . . . .43
    Tanks Dimensions and
        Capacities. . . . . . . . . . . . . . . .44
    Construction Features . . . . . . . . . .44
Tank Container Markings. . . . . . . . . . . .47
Types of Nonpressure Intermodal Tank
    Containers . . . . . . . . . . . . . . . . . . .51
    Nonpressure Intermodal Tank
        Containers . . . . . . . . . . . . . . .51
    Tank Container Safety
        Devices . . . . . . . . . . . . . . . . .57
Inspections and Testing. . . . . . . . . . . .58
    Inspections . . . . . . . . . . . . . . . . .58
    Tank and Valve Testing . . . . . . . . .58
Summary. . . . . . . . . . . . . . . . . . . . . . .58
References . . . . . . . . . . . . . . . . . . . . . .58

**CHAPTER 6 Intermodal Tank Containers for Gases . . . . . . . . . . . . . . . . . . . . . 59**

Chapter Outline . . . . . . . . . . . . . . . . . .59
Key Terms . . . . . . . . . . . . . . . . . . . . . .59
Introduction . . . . . . . . . . . . . . . . . . . .59
Pressure Tank Containers . . . . . . . . . . .59
    Pressure Tank Container
        Features . . . . . . . . . . . . . . . . .60
    Data Plate . . . . . . . . . . . . . . . . .60
    Pressure Tank Container
        Fittings . . . . . . . . . . . . . . . . .61
    Pressure Tank Container Safety
        Devices . . . . . . . . . . . . . . . . .62
Cryogenic Tank Containers . . . . . . . . . .63
    Holding Time. . . . . . . . . . . . . . . .64
    Features. . . . . . . . . . . . . . . . . . .64
Multiple Element Gas Containers
    (MEGCs) . . . . . . . . . . . . . . . . . . .65

Summary. . . . . . . . . . . . . . . . . . . . . . .65
References . . . . . . . . . . . . . . . . . . . . . .67

**CHAPTER 7 Intermodal Container Operations . . 68**

Chapter Outline . . . . . . . . . . . . . . . . . .68
Key Terms . . . . . . . . . . . . . . . . . . . . . .68
Introduction . . . . . . . . . . . . . . . . . . . .68
Highway Operations. . . . . . . . . . . . . . .68
    Highway Equipment . . . . . . . . . . .69
Railroad Operations . . . . . . . . . . . . . . .70
    Equipment. . . . . . . . . . . . . . . . . .70
    COFCs. . . . . . . . . . . . . . . . . . . . .70
    TOFCs. . . . . . . . . . . . . . . . . . . . .70
    Intermodal Portable Tank
        Containers . . . . . . . . . . . . . . .71
Marine Operations . . . . . . . . . . . . . . . .72
    Types of Intermodal Container
        Ships. . . . . . . . . . . . . . . . . . .73
    Container Ship Cargo Handling
        and Stowage . . . . . . . . . . . . .75
Fixed Facility Operations . . . . . . . . . . .75
    Temporary Storage . . . . . . . . . . . .75
    Permanent and Semipermanent
        Storage . . . . . . . . . . . . . . . . .76
    Loading and Off-Loading
        Methods . . . . . . . . . . . . . . . .77
    Loading Methods . . . . . . . . . . . . .77
    Off-Loading Methods . . . . . . . . . .77
    Safety Guidelines for Transfer
        Operations . . . . . . . . . . . . . . .78
Summary. . . . . . . . . . . . . . . . . . . . . . .80
References . . . . . . . . . . . . . . . . . . . . . .80

**CHAPTER 8 Emergency Response . . . . . . . . . . . 81**

Chapter Outline . . . . . . . . . . . . . . . . . .81
Key Terms . . . . . . . . . . . . . . . . . . . . . .81
Introduction . . . . . . . . . . . . . . . . . . . .81
Intermodal Portable Tank Container
    Emergencies. . . . . . . . . . . . . . . . . .81
    Damage Assessment . . . . . . . . . . .81
    IM-101, IM-102 (International—IMO
        Type 1 and 2 – T Code T-1 – T-22)
        Leak Control . . . . . . . . . . . . . .82
Intermodal Freight Container
    Emergencies. . . . . . . . . . . . . . . . . .84
Industrial Facility Emergencies . . . . . . .87
Container Terminal Emergencies. . . . . . .89
Highway Transportation Emergencies . .90
Railroad Transportation Emergencies . .91
    Trailer On A Flat Car and Container
        On A Flat Car. . . . . . . . . . . . . .91
Marine Transportation Emergencies. . . .92
Summary. . . . . . . . . . . . . . . . . . . . . . .96
Reference . . . . . . . . . . . . . . . . . . . . . .96

**CHAPTER 9  Uprighting, Product Removal, and Transfer Operations** . . . . . . . . . . . . **97**

Chapter Outline . . . . . . . . . . . . . . . . . . . .97
Key Terms . . . . . . . . . . . . . . . . . . . . . . . .97
Introduction . . . . . . . . . . . . . . . . . . . . . .97
Uprighting, Removal, and Transfer
    Operations . . . . . . . . . . . . . . . . . . . . .97
      Surveying the
        Container . . . . . . . . . . . . . . . . . .97
Site Safety Measures . . . . . . . . . . . . . . .98
Grounding and Bonding
    Considerations . . . . . . . . . . . . . . . . . .98

Product Transfer Methods . . . . . . . . . . . .99
Uprighting Methods . . . . . . . . . . . . . . . .101
Summary . . . . . . . . . . . . . . . . . . . . . . . .104
References . . . . . . . . . . . . . . . . . . . . . . .104

**APPENDIX**
  **Acronyms** . . . . . . . . . . . . . . . . . . . . . . . . . . . . . . . . **105**

Glossary . . . . . . . . . . . . . . . . . . . . . . . . . . . . . . . . . . . .106

Index . . . . . . . . . . . . . . . . . . . . . . . . . . . . . . . . . . . . . 110

# Acknowledgments

Courtesy of Bill Hand.

The second edition of *Intermodal Container Emergencies* is designed to assist the user in meeting the knowledge requirements outlined in NFPA 472, Standard for Competence of Responders to Hazardous Materials/Weapons of Mass Destruction Incidents, Chapter 14 - Competencies for Hazardous Materials Technicians with a Intermodal Tank Specialty.

The terminology used throughout this text is based on the terms adopted by the United Nations, *Recommendations on the Transport of Dangerous Goods—Model Regulations*, Nineteenth revised edition, New York and Geneva. More information on UN regulations is available in Chapter 2.

It should be emphasized that reading this publication will not provide the reader with all of the necessary knowledge to respond to a hazardous materials incident. For more information consult the textbook, *Hazardous Materials: Managing the Incident*, 4th edition (2014) by Jones & Bartlett Learning, LLC.

No author gets it right the first time, and over the years we have found the third-party review process to result in a much better final work product. Throughout the development process we relied on our friend and HazMat colleague Charles J. Wright to be our technical reviewer on some topics and the Glossary. Intermodal containers and the regulations and standards that govern them are very complex. Charlie did an excellent job offering suggestions on how to improve terminology and arrange the manuscript so it is easier for the reader to better understand the material. Charlie also willingly shared some of his on materials and references. Our text clearly would not have been as good as it is without his input.

We want to acknowledge contributions made by the following people for reviewing the final page proofs for accuracy and for sharing their written materials, photographs, and ideas for improving the book.

Captain Michael Oder, Port of Houston (TX) Marine Fire Department (Retired)

Mr. Stephen D. McConachie, U.S. Customs and Border Protection

Finally, as authors we tried to tackle the complicated topic of intermodal containers and simplify as much as possible. While it is true that intermodal containers are standardized, the various fittings, attachments, and valves vary a great deal because manufacturers provide many options for the customer and some containers are customized. Every time we visit a container yard or ship we see and learn something new. There is no practical way we could cover every type of container variation in this book. Many intermodal container and valve manufacturers have web sites with very useful technical information on their products and specifications. This information is usually current and easily accessed using a search engine.

## About the Authors

MICHAEL HILDEBRAND and GREGORY NOLL have more than 40 years of emergency preparedness and response experience. They have served as firefighters, hazardous materials technicians, instructors, incident commanders, and on national level Incident Management Teams. They are both United States Air Force veteran firefighters.

BILL HAND is a United States Army veteran and is retired from the Houston Fire Department after 31 years of service with the Hazardous Materials Response Team. During his career, he responded to over 7,500 hazmat incidents. Bill has served on numerous technical committees and reviewed and developed HazMat training programs for over 40 years.

All three authors are recipients of the International Association of Fire Chiefs (IAFC) John M. Eversole Lifetime Achievement Award for their leadership and contributions to the hazardous materials emergency response community.

Courtesy of Tyler Bones

# Introduction

## Chapter Outline

- Key Terms
- Introduction
- Target Audience
- Overview of the Intermodal Container Industry
- Objectives
- Summary
- References

## Key Terms

**Containerization** A system of intermodal freight transportation that uses standard containers that can be loaded onto vessels, railcars, and trucks. It involves the stowage of general or special cargo in a container for transport in the various modes.

**Intermodal Container** A standardized shipping container built to transport materials in multiple modes—road, rail, water, and air—without unloading and reloading the cargo, fitted with a rigid frame with corner castings for tie-down and lifting to facilitate mechanical handling with its contents intact.

**Intermodal Portable Tank Container** An intermodal container for transporting liquids, solids, and gases in bulk with a capacity of 118.9 gallons (450 L) or more. See the glossary for a more detailed definition.

**Freight Container** An intermodal container for transporting packages in unit form with a volume of 64 cubic feet (1.8 cubic meters) or more intended primarily for containment of packages during transportation. See the glossary for a more detailed definition.

**Multiple Element Gas Container (MEGC)** Assemblies of UN high pressure cylinders, tubes, or bundles of cylinders interconnected by a manifold and assembled within a rigid frame with corner castings for the transport of gases. Also known as a tube module.

**Tank Container** An intermodal container for transporting liquids, solids, and gases in bulk with a capacity of 118.9 gallons (450 L) or more and includes Multiple Element Gas Containers (MEGC), also called tube modules, which are the high-pressure equivalent of tank containers.

## Introduction

The use of intermodal containers (freight and tank) and the companies that transport intermodal containers are vital to both the national and international economy and security. It is a multibillion dollar industry that continues to grow globally.

The U.S. marine, rail, and highway transportation systems handle large volumes of domestic and international freight in support of the nation's economic activities. As a vital part of that system, the nation's container ports and terminals handle large volumes of intermodal container cargo and are sources of employment, revenue, and taxes for businesses and the communities where they are located.

Modern intermodal containers carry numerous commodities; if something is grown or manufactured anywhere in the world, it is probably shipped in an intermodal container. Containers have reduced the cost of safely transporting such goods as children's toys, clothing, and electronics from factories halfway across the globe to neighborhood discount stores across America, and cut the time it takes to load and unload the large vessels used in transporting these goods. Similarly, tons of frozen meats and manufactured machinery and parts from all across the United States are regularly shipped in containers to markets abroad.

Almost any Hazard Class may be found in an intermodal container. Hazardous and nonhazardous materials are shipped in specialized intermodal tank containers. Many intermodal freight containers also carry a variety of hazardous materials packed inside them in smaller packages, boxes, crates, drums, or cylinders.

When an accident or fire occurs involving these containers they can present special and unique problems for emergency responders and carriers, especially when they are on board a ship at sea.

In addition to the potential for accidents and fires, the international nature of the intermodal container business presents certain security risks to the United States. Containers have been used by drug smugglers, gunrunners, and human traffickers for illegal purposes. In today's threat environment, there is a risk of intermodal containers being used by terrorist groups. Securing our ships at sea and our homeland ports is both complicated and challenging.

Dealing with intermodal container emergencies is a team effort that requires emergency responders, law enforcement, shippers, carriers, and product and container specialists to share information and cooperate to bring the incident to safe closure.

## Target Audience

The target audience for this text is primarily public safety and industrial emergency responders, law enforcement and port security officers, customs and inspection officers, and hazardous materials cargo recovery and environmental contractors. Other groups that may find various chapters useful include shippers, longshoremen and stevedores, merchant mariners, and insurance underwriters.

This publication is designed to provide background information on intermodal container design and construction features. We will also provide an overview of intermodal container operations for sea, land, and rail, as well as guidance for handling response issues specific to each mode of transportation. The text will also cover various types of shipping containers and packaging including portable tank and other nonbulk portable and bulk packaging commonly encountered inside of intermodal freight containers.

## Overview of the Intermodal Container Industry

### ■ Purpose of an Intermodal Container

The purpose of an intermodal container is simple—to have a transport vehicle that can be physically moved from one location to another on more than one mode of transportation, e.g., from ship to trucks or railcars and back to ships without unloading the container's contents.

Intermodal containers are sometimes shipped by commercial or military aircraft. Movement of containers by commercial or military cargo aircraft usually involves special high value or urgent cargo. The U.S. Department of Defense Transportation Regulation, Part VI, Chapter 607, *Movement of Containers by Air* (March 2016) provides guidance on transporting 20-foot intermodal freight containers.

The primary reason the containers are "intermodal" is because freight and portable tank containers are constructed to standard international designs. This factor makes the intermodal shipping industry work efficiently and cost effectively. Without standardization, the system would fail. With globalization of the international economy, standards are crucial to facilitate the efficient transfer of cargo between the various transportation modes. The standardization process has been facilitated through codes created by the United Nations and then adopted by member nations.

The typical intermodal freight container is 20 or 40 feet in length. The freight containers are commonly known as a "TEU" (20-foot Total Equivalent Unit) or as an "FEU" (40-foot Total Equivalent Unit). They also have standard heights and widths. (See **FIGURE 1-1**.) Standardization allows containers to transit international borders using a variety of cargo loading and transportation platforms (e.g., the modern container ship). Some special purpose containers may have different dimensions.

Intermodal containers are the backbone of the American cargo industry. According to a U.S. Customs and Border Protection (CBP) 2013 report, approximately 25 million intermodal containers entered the United States.

Most intermodal containers that arrive or depart the United States move through marine terminals where they are loaded onto trucks or rail cars for the journey to their destination. It is not unusual for a container to move through several modes of transportation on its journey.

Looking back at history, 60 years ago things were very different. Prior to 1957, the majority of bulk cargo in foreign trade moved over the world's oceans in break bulk cargo ships. Loading cargo was a laborious and time-consuming process that involved the hard work of stevedores and longshoremen. In those days the schedules of steamships were uncertain; consequently, outbound cargo was often delivered to the pier days or weeks ahead of the ships' arrival in port.

The typical cargo ship carried a wide range of goods that would vary from food and clothing to building materials or heavy machinery. All of this cargo needed to be moved from the warehouse to the pier, and then lifted and loaded onto the ship. Once on board, the cargo was stowed below deck in the cargo

**FIGURE 1-1**  Intermodal containers are standardized, which allows for easy shipment in various modes of transportation.

Courtesy of Bill Hand.

hold to ensure it would arrive undamaged at its destination. Some of the cargo would need to be braced and secured with wood or matting. On the other end of the journey, the ship needed to be unloaded and then all that cargo would be once again loaded onto trucks or rail cars in order to get to its final destination. It was a bewildering and expensive process. It was also very dangerous work. (See **FIGURE 1-2**.)

Work on standardizing containers began in the 1940s. In 1956, the owner of a North Carolina based trucking company (Malcolm McLean) came up with the idea to place cargo inside a box similar to a tractor-trailer rig. The general idea was to load the container with its cargo and then lift the entire container onto the deck of the ship. Today this idea sounds simple, but prior to 1956 it had never been done before.

Mr. McLean's biggest challenge was the lack of any ships designed to handle containers either above or below deck. To solve the problem, he turned to oil

**FIGURE 1-2**  Prior to 1956 all dry bulk cargo moved over the world's oceans in a cargo ship.

Courtesy of The Mariners' Museum.

tankers. The typical oil tanker offloads its petroleum liquid cargo and returns for another load with its cargo tanks empty. Ballast is added to stabilize the ship at sea by pumping water into the cargo tanks. McLean's idea was to modify an oil tanker by adding a cargo deck on the stern and trade water weight for ballast by placing loaded containers on the deck. He installed a spar deck on a modified tanker and designed longitudinal slots, which held the trailer trucks in place during transit. Eventually the trailer trucks evolved into what we now call an intermodal freight container.

On April 26, 1956, McLean's experimental oil tanker/container ship called the *Ideal X* departed Newark, New Jersey with 56 trailer trucks for Houston, Texas. From that day forward, the shipping industry changed forever.

## ■ The Land Bridge Concept

The biggest advantage of intermodal container shipping is that the cargo is typically loaded at the factory, sealed, and then securely loaded on a truck or rail car. Once the container arrives at a port, it remains sealed and secured until it is ready for loading onto the container ship. Bar coding, modern container movers, stackers, and cranes allow ships to be loaded and unloaded rapidly. The entire operation is very fluid. (See **FIGURE 1-3** .)

Use of intermodal containers within North America rose during the 1970s when the United States began to be used as a "land bridge" for international traffic between the Atlantic and Pacific Oceans and the Gulf of Mexico. Under this service, offered by the railroads to various ocean shipping companies, a container initially loaded in Europe and en route to the Far East would travel via container ship to the U.S. East Coast, be loaded onto a train for the trip across the continental United States, and then be reloaded onto another container ship on the U.S. West Coast for its ultimate destination in Asia. (See **FIGURE 1-4** .)

In the 1980s, the use of intermodal containers became even more prevalent for domestic transportation, particularly for intermodal freight containers and their cousin the piggyback trailer.

Over the last 50 years, container ships have grown in both size and cargo capability. In 1956, McLean's first container ship carried just 56 containers. Today, the 100 largest container ships in the world built between 2006 and 2016 have capacities between 13,000 and 19,000 TEUs! (See **FIGURE 1.5** .)

Today, 90 percent of nonbulk cargo is moved by container ships. A single container ship can carry the equivalent of 30 freight trains one-mile long stacked two containers high. One of the largest container ships in the world, the *MSC Oscar*, is 1,297 feet in length and can take on 19,224 containers. To put that into perspective, consider that the U.S. Navy's Nimitz Class aircraft carriers are about 1,300 feet in length.

## Objectives

This text is designed to meet the general requirements for handling intermodal container emergencies as described in National Fire Protection Association Standard NFPA© 472, *Competence of Responders to Hazardous Materials/Weapons of Mass Destruction Incidents (2013)*, Chapter 7—Hazardous Materials Technician. Certain sections of the

**FIGURE 1-3** Intermodal container shipping is both time and cost efficient.

Courtesy of Bill Hand.

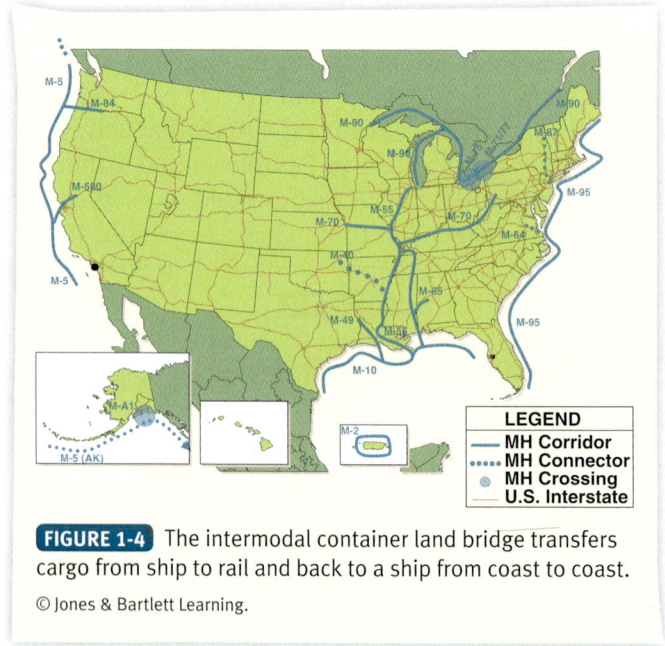

**FIGURE 1-4** The intermodal container land bridge transfers cargo from ship to rail and back to a ship from coast to coast.

© Jones & Bartlett Learning.

**FIGURE 1-5** Modern container ships can handle between 13,000 and 19,000 TEUs.

Courtesy of Bill Hand.

textbook will also meet the requirements of NFPA© 472, Chapter 14—*Hazardous Materials Technicians with an Intermodal Tank Specialty*. Text objectives that satisfy NFPA 472 requirements are noted, as appropriate. There are also some basic intermodal competencies in Chapters 4 and 5.

The terminology used throughout this text is based on terms adopted by the United Nations, 2015 Recommendations on the Transport of Dangerous Goods - Model Regulations, 19th Edition, New York, NY (2015). More information on UN regulations is available in Chapter 2.

For more information concerning meeting the broader Hazardous Materials Technician knowledge and skills objectives of NFPA© 472, consult the text, *Hazardous Materials, Managing the Incident*, 4th edition, Jones and Bartlett Learning.

1. Describe the major codes, regulations, and/or standards which govern the transportation and use of intermodal containers (freight containers and tank containers), including:
   a) International Maritime Organization (IMO)
   b) U.S. Department of Transportation (DOT)
   c) Transport Canada (TC)
   d) Association of American Railroads, Specification for the Acceptability of Tank Containers (AAR 600)
   e) International Standards Organization (ISO)
   f) UN Recommendations on the Transport of Dangerous Goods—Model Regulations (19th Edition) UNECE

2. Identify the following tank containers by name and specification, their typical contents by name and hazard class, and the basic design and purpose of tank components, including corner casting, data plate, heater coils, insulation, jacket, refrigeration unit, and supporting frame. [NFPA 472 – 7.2.1.1.2, NFPA 472 – 7.2.3.1, NFPA – 7.2.3.1.1. NFPA 472 – 14.2.1.1, NFPA 14.2.1.2, NFPA – 14.2.1.3]

3. Nonpressure tank containers (UN portable tank T-11–T-22)
   a) IM-101 portable tanks (US and Canada)
   b) IM-102 portable tanks (US and Canada)
   c) IMO Type 1 (international)
   d) IMO Type 2 (international)

4. Pressure tank containers
   a) UN portable tank containers (T50) (international)
   b) DOT Specification 51 (US and Canada)
   c) IMO Type 5 (international)

5. Cryogenic liquid intermodal tank containers
   a) UN portable tank containers (T75) (international)
   b) IMO Type 7 (international)

6. Multiple Element Gas Containers (MEGCs) and/or Tube Modules
   a) Multiple element gas containers (MEGCs) (international)
   b) IMO Type 7 (international)

7. Given examples of various fittings arrangements for pressure, nonpressure, and cryogenic intermodal tank containers, identify and describe the design, construction, and operation of each of the following fittings, where present [NFPA 472 – 7.2.3.1.1. (4), NFPA – 14.2.1.4]:
   a) Airline connection
   b) Bottom outlet valve
   c) Gauging device
   d) Liquid or vapor valve
   e) Thermometer
   f) Manhole cover
   g) Pressure gauge
   h) Sample valve
   i) Spill box
   j) Thermometer well
   k) Nonpressure tank container airline connection
   l) Pressure tank container vapor valve

8. Given examples of various safety devices for pressure, nonpressure, and cryogenic intermodal tank containers, identify and describe the design, construction, and operation of each of the following safety devices, where present [NFPA – 14.2.1. (5)]:
   a) Emergency remote shutoff device
   b) Excess flow valve

c) Fusible link/nut assemblies

d) Regulator valve

e) Rupture disc

f) Pressure and nonpressure tank containers

g) Pressure relief valve on pressurized tank containers

9. Describe the construction features and container markings on an intermodal freight container, including the following [NFPA 472 – 5.2.1.2.1]:

a) Container identification (owner's code, category identified (U for intermodal), serial number, and check digit)

b) Country, size, and type markings (when required)

10. Describe the container markings on an intermodal portable tank container, including the following [NFPA 472 – 5.2.1.2.1]:

a) Reporting marks and number

b) Domestic and international specification marking

c) DOT special approval marking

d) AAR-600 marking

e) Country, size, and type markings

f) Special hazards and cargo handling restrictions marking

11. Given leaks from the following fittings on intermodal tanks, control the leaks using approved methods and procedures [NFPA – 14.4.1]:

a) Bottom outlet

b) Liquid/vapor valve

c) Manhole cover

d) Pressure relief device

e) Tank

12. Given a pressure intermodal portable tank container containing a liquefied gas, describe the methods for determining the amount of liquid in the tank [NFPA 472 – 14.2.1.9].

13. Given an intermodal portable tank involved in an emergency, identify the factors to be evaluated as part of the intermodal tank damage assessment process, including the following [NFPA – 1.2.1.7, 14.2.1.8, 14.2.2.2]:

a) Amount of product released and amount remaining in the intermodal portable tank container

b) Container stress applied to the intermodal portable tank

c) Stress applied to the container frame (box, beam)

d) Types of container construction materials

e) Nature of the emergency

f) Number of compartments

g) Pressurized or nonpressurized

h) Type and nature of tank damage

i) Type of intermodal tank

j) Type of tank metal

13. Describe the likely breach/release mechanism for each of the following types of intermodal tank containers:

a) UN portable tank container T-11–T-22, T-50 and T-75

b) IMO Type 1/ IMO-101

c) IMO Type 2/ IMO-102

d) IMO Type 5/ DOT-51

e) DOT-56

f) DOT-57

g) DOT-60

h) Cryogenic (IMO Type 7)

14. Given an emergency involving an intermodal freight container, describe procedures for the following:

a) Safely opening intermodal freight container doors in an upright or overturned accident condition

b) Conducting forcible entry into an intermodal freight container through damaged freight container doors

c) Fire tactics for extinguishing an interior fire on an intermodal freight container

d) Hazardous materials release from nonbulk and intermediate bulk containers within the intermodal freight container

15. Describe how cargo is stored for safe transportation inside freight containers and the most common types of containers shipped that present safety, fire, or hazardous materials emergency response issues, including the following:

a) Drums

b) Bags

c) Boxes

d) Intermediate Bulk Containers (IBC)

16. Describe the methods for containing the following leaks on nonpressure liquid intermodal tanks (e.g., T Code T-1–T-22, IM-101 and IM-102) [NFPA – 14.4.5]:

a) Manhole cover leak

b) Irregular-shaped hole

c) Pressure relief devices (e.g., vents, rupture disc)

d) Puncture

e) Split or tear

f) Valves and piping

17. Describe the methods for containing the following leaks in pressurized intermodal tank containers (e.g., T Code T50, Specification 51, IMO Type 5) [NFPA 472 – 14.4.6]:

a) Crack

b) Failure of pressure relief device (e.g., relief valve, rupture disc)

c) Valves and piping

18. Describe the normal operations, handling procedures, and emergency response concerns for intermodal containers in each of the following situations, including:
    - Terminals
    - Highway transportation
    - Railroad transportation
    - Marine transportation
    - Fixed facilities
19. Describe the bonding and grounding procedures for product transfer from an intermodal portable tank container to a recovery container [NFPA 472 – 14.4.4].
20. Describe the application and use of the following product transfer and recovery equipment for an emergency involving a portable tank container:

    a) Portable pumps (air, electrical, pneumatic)
    b) Pressure transfers
    c) Vacuum trucks
    d) Vehicles with power-takeoff driven pumps

## Summary

Containerization and the companies that transport intermodal containers are vital to our nation's economy and security. It is a multibillion dollar industry that continues to grow globally.

Almost any DOT Hazard Class may be found in an intermodal container. Hazardous materials are shipped in specialized intermodal tank containers. Many intermodal freight containers also carry a variety of hazardous materials packed inside them in smaller packages, boxes, or drums. The target audience for this text is primarily public safety and industrial emergency responders, law enforcement and port security officers, customs and inspection officers, and hazardous materials cargo recovery and environmental contractors.

The purpose of an intermodal container is simple—to have a transport vehicle that can be physically moved from one location to another on more than one mode of transportation, e.g., from ship to trucks or railcars and back to ships.

Most intermodal containers that arrive or depart the United States move through marine terminals where they are loaded onto trucks or rail cars for the journey to their destination. It is not unusual for a container to move through several modes of transportation on its journey.

This text is designed to meet the general requirements for handling intermodal container emergencies as described in National Fire Protection Association standard NFPA© 472, *Competence of Responders to Hazardous Materials/Weapons of Mass Destruction Incidents (2013)*, Chapter 7—Hazardous Materials Technician. Certain sections of the text will also meet the requirements of NFPA© 472, Chapter 14—*Hazardous Materials Technicians with an Intermodal Tank Specialty*.

## References

1. Cudahy, Brian J. 2006. "The Containership Revolution: Malcom McLean's 1956 Innovation Goes Global." *TR News*. (September-October), pp. 10–12.
2. Harrison, Robert. 2006. "Can Intermodal Freight Terminals Handle Supersizing?" *TR News*. (September-October), pp. 13–17.
3. International Tank Container Organisation. 2013. *Tank Container Fleet Survey*. Romford, Essex, United Kingdom.
4. Noll, Gregory, G., and Michael S. Hildebrand, with contributions by Glen Rudner and Rob Schnepp. 2014. *Hazardous Materials: Managing the Incident*. 4th. ed. Burlington, MA: Jones & Bartlett Learning.
5. Katims, Ron. 2006. "The 40-Foot Container: Industry Standard Faces Challenges and Changes." *TR News*. (September-October), pp. 18–23.
6. Tailor, L.G., Captain and Captain L.D. Conway. 1992. *Cargo Work: The Care, Handling and Carriage of Cargos*. Glasgow, Scotland: Brown, Son & Ferguson.
7. Tomlinson, John. 2005. "History and Impact of the Intermodal Shipping Container." *LIS 654-05/Carrie Bickner*. (September 22), pp. 1–7.
8. United Nations. *2015 Recommendations on the Transport of Dangerous Goods - Model Regulations*. 19th ed. New York.
9. U.S. Department of Transportation, Bureau of Transportation Statistics. January 2011. *America's Home Container Ports: Linking Markets at Home and Abroad*, p. 28.

# Codes, Standards, and Regulations

## Chapter Outline

- Key Terms
- Introduction
- United Nations Recommendations on the Transport of Dangerous Goods—Model Regulations
- International Maritime Dangerous Goods (IMDG) Code
- U.S. Department of Transportation (U.S. DOT)
- Transport Canada (TC)
- Normas Oficiales Mexicanas (NOM)
- Association of American Railroads (AAR)
- International Standards Organization (ISO)
- Convention for Safe Containers (CSC)
- Transportation Security Regulations
- Summary
- References

## Key Terms

**Dangerous Goods** As defined by the United Nations and the International Maritime Dangerous Goods (IMDG) Code, these are materials or items with hazardous properties which, if not properly controlled, present a potential hazard to human health and safety, infrastructure, and/or their means of transport.

**Hazardous Materials** As defined by the U.S. DOT [49 CFR 171.8], any substance or material capable of posing an unreasonable risk to health, safety, and property when transported in commerce. This includes hazardous substances, hazardous wastes, marine pollutants, and elevated temperature materials.

**Transportation Worker Identification Credential (TWIC®)** The Transportation Worker Identification Credential (TWIC) is a security program designed to ensure that individuals who pose a threat do not enter a marine terminal, a facility attached to a marine terminal, or a merchant marine vessel. The TWIC is issued by the U.S. Department of Homeland Security (DHS). All licensed merchant mariners and officers are required to have a TWIC.

## Introduction

Intermodal containers are designed, constructed, and handled in accordance with a number of domestic and international codes, regulations, and consensus standards. The objective of this chapter is to provide an overview of the key U.S., international, and foreign domestic standards that influence both normal handling and emergency response operations at incidents involving intermodal containers.

Depending upon one's role in the transportation lifecycle, codes, standards, and regulations are used differently. For example, shippers, inspectors, and law enforcement will concentrate on issues that are distinctly different from emergency responders. The focus of this chapter is to outline the key regulations that are referenced in the intermodal container industry, the primary agencies and players who may be involved in regulatory enforcement, and how additional information can be accessed.

## United Nations Recommendations on the Transport of Dangerous Goods—Model Regulations

The 19th edition of the UN's Recommendations on the Transport of Dangerous Goods addresses governments and international organizations concerned with safety in the transport of dangerous goods. This publication has been prepared by the secretariat of the United Nations Economic Commission for Europe (UNECE) which provides secretariat services to the Economic and Social Council's Committee of Experts. These recommendations are significant for the following reasons:

1. The recommendations have been developed by the United Nations experts on the Transport of Dangerous Goods. The recommendations are addressed to governments and international organizations concerned with the regulation of the transport of dangerous goods. They do not apply to the bulk transport of dangerous goods in sea-going or inland navigation bulk carriers or tank vessels, which is subject to special international or national regulations.

2. The recommendations concerning the transport of dangerous goods are presented in the form of "Model Regulations on the Transport of Dangerous Goods," as an annex to the document. Some governments have adopted the Model Regulations as law; for example, the United States has adopted model regulations through the U.S. Department of Transportation. However, the Model Regulations are flexible enough to accommodate any special requirements that might have to be met by an adopting government or agency. The intent of Model Regulations is to facilitate worldwide harmonization in trade, which lowers cost and improves safety.

3. The scope of the Model Regulations covers principles of classification and definition of classes, listing of the principal dangerous goods, general packing requirements, testing procedures, marking, labeling or placarding, and transport documents. There are special requirements related to particular classes of goods. With this system of classification, listing, packing, marking, labeling, placarding, and documentation in general use, the carriers, consignors, and inspecting authorities benefit from simplified transport, handling, and control and from a reduction in time-consuming formalities.

4. The Model Regulations have been adopted by U.S. Department of Transportation, and are incorporated into the IMDG Code.

## International Maritime Dangerous Goods (IMDG) Code

The primary intermodal container reference document for the transportation of dangerous goods and intermodal containers is the International Maritime Dangerous Goods (IMDG) Code. The IMDG Code is developed and updated bi-annually through the efforts of the International Maritime Organization (IMO), which is a United Nations specialized agency with global standard-setting authority for the safety, security, and environmental performance of international shipping.

The objective of the IMDG Code is to enhance the safe transport of dangerous goods, protect the marine environment, and facilitate the free unrestricted movement of dangerous goods. It is intended for use not only by the mariner but also by all those involved in industries and services connected with shipping, and contains advice on terminology, packaging, labeling, placarding, markings, stowage, segregation, handling, and emergency response action.

Although the IMDG Code by itself does not have the force of law, the Code is required to be adopted by the International Convention for the Safety of Life at Sea (SOLAS) and the International Convention for the Prevention of Pollution from Ships (MARPOL 73/78) signatory states. As a result, it becomes the de facto international standard. (See **FIGURE 2-1**.)

**FIGURE 2-1** The design, construction, movement, and handling of intermodal freight and tank containers are governed through a network of international, internal, and industry standards.

Courtesy of Greg Noll.

## U.S. Department of Transportation (U.S. DOT)

Within the U.S. Department of Transportation (DOT), the Pipeline and Hazardous Materials Safety Administration (PHMSA) is charged with the responsibility of protecting people and the environment from the risks of hazardous materials transportation. Given the multimodal nature of intermodal freight and portable tank containers and the movement of hazardous materials, PHMSA has regulatory responsibility for the highway and rail modes of transport, while the U.S. Coast Guard has responsibility for marine transport. The Federal Motor Carrier Safety Administration (FMCSA) may also be involved in evaluating the safety fitness of motor carriers transporting intermodal containers via highway.

Title 49 of the Code of Federal Regulations (49 CFR) covers the transportation of hazardous materials. Parts 170 through 179 define the types of portable intermodal tanks, design and construction requirements, special product-related features, and inspection requirements. Part 180 addresses continuing qualification and maintenance of packaging and is the basis for grandfathering in older tank containers under the United Nations Regulations recommendations.

Any deviations from the DOT requirements require a Special Permit approval. If a portable intermodal tank container is issued a Special Permit, the number will be stenciled on the side of the container (e.g., DOT-SPXXXX). Among the reasons for Special Permits are to:

- Authorize shipment of a regulated material in DOT-approved packaging not currently authorized.
- Develop information about and gain experience concerning the unique forms of containerization, shipping conditions, and carrier operations.
- Authorize packaging of a nature and integrity equivalent to DOT specified containers.
- Allow single trip or emergency movement of hazardous materials for which an amendment or the DOT regulations are impractical.

In the past there were some differences between the DOT and IMDG requirements, but many of them have been minimized by the UN Recommendations. Some examples include shell thickness calculations and pressure relief valve settings.

## Transport Canada (TC)

Transport Canada is the equivalent Canadian agency of the U.S. Department of Transportation. The TC Transport of Dangerous Goods Directorate (TDG) serves as the major source of regulatory development, information, and guidance on dangerous good transport for all modes of transportation. All Canadian provinces have adopted the UN's Recommendations on the Transport of Dangerous Goods. Because of the high number of cross-border shipments, there is considerable reciprocity between the United States and Canada in adopting the UN Recommendations as regulations.

For certain higher risk dangerous goods, the TDG regulators may require the shipper or the person offering the dangerous good for shipment to develop an Emergency Response Assistance Plan (ERAP). The ERAP is intended to assist local emergency responders by providing them with technical specialists and specially trained and equipped emergency response personnel at the scene of an incident.

## Normas Oficiales Mexicanas (NOM)

The Official Mexican Standards (Normas Oficiales Mexicanas or NOMs) augment the Mexican Regulation for the Land Transport of Hazardous Materials and Wastes. The Mexican Secretariat for Communications and Transport (SCT) is responsible for publishing and maintaining the NOMs. In addition, other Mexican government agencies have published standards relevant to the transportation of hazardous materials within Mexico. The Mexican hazardous materials NOMs are fairly consistent with those of the United Nations Recommendations on the Transport of Dangerous Goods. Since the U.S. Hazardous Materials Regulations are also periodically harmonized with the UN Model Regulations, the requirements of the hazardous materials regulations and the Mexican NOMs are fairly consistent, which facilitates international trade between the two countries. However, there are some differences, which will continue to be addressed by the Land Transportation Standards Hazardous Materials Working Group (LTSS Group 5) and through amendments and restructuring of the UN Recommendations. English translations of the regulations may be accessed through any web-based search engine such as Google.

## Association of American Railroads (AAR)

The Association of American Railroads (AAR) is a U.S. trade association for the railroad industry which also develops industry standards and guidelines. As the standards development organization for North American railroads, AAR also conducts research and testing through its

Transportation Technology Center Inc. (TTC) in Pueblo, Colorado. The TTC facility also includes a world-class hazardous materials training center that focuses on railroad emergency scenarios, including intermodal tank and freight containers.

All portable tank containers which are accepted for transport on the railroad must meet the requirements of AAR 600, "Specification for Acceptability of Tank Containers." All tank containers which meet these requirements are stenciled with AAR 600 on each side of the container. U.S. DOT/PHMSA has also adopted AAR 600 into its hazardous materials regulations. (See **FIGURE 2-2**.)

## International Standards Organization (ISO)

ISO standards cover the structural design of intermodal containers, thereby allowing for worldwide intermodal movement. ISO standards provide specifications for the manufacture of intermodal containers, including size, strength, and durability, and cover the corner casting design, dimensional design, racking and stacking requirements, and markings relative to size, strength, and identification. Although ISO is not law, it does establish the standards under which ships, trailers and chassis, and rail cars are used for intermodal transportation requirements.

Among the most used ISO standards that impact intermodal containers and shipments is ISO 6346—Container Codes. ISO 6346 covers the coding, identification, and marking of intermodal (shipping)

Example of an ISO 6346 conform container number:

# CSQU3054383

**Owner code** → **Category identifier**    **Serial number** → **Check digit**

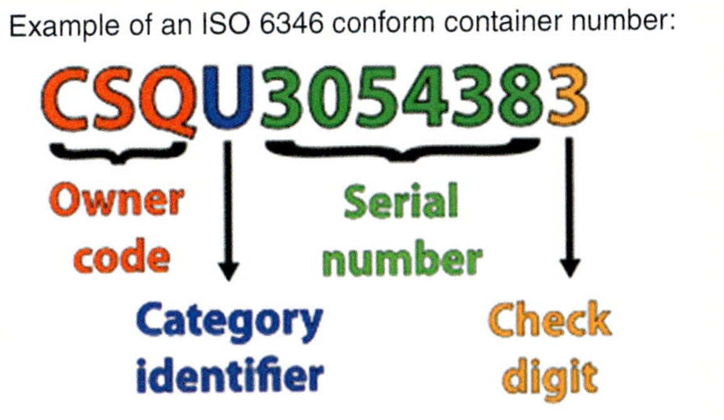

**FIGURE 2-3** Example of an ISO 6346 compliant container number.

© Jones & Bartlett Learning.

containers used within containerized intermodal freight transport. (See **FIGURE 2-3**.) Managed by the International Container Bureau (BIC), the standard establishes a visual identification system for every container that includes a unique serial number (with check digit), the owner, a country code, a size, type, and equipment category as well as any operational marks.

## Convention for Safe Containers (CSC)

CSC is an international convention that is primarily intended to ensure that all intermodal containers (i.e., freight, tank, etc.) undergo regular inspections for structural integrity. Certain strength, testing, and markings are included in the convention. CSC is referenced by DOT in 49 CFR Parts 450 through 453, and compliance is required by 49 CFR Part 173.32 b (c). The current CSC was published in 2012, and is updated on approximately 10-year cycles.

ISO Container Safety Certificates (CSC Plate) are issued by the container manufacturer and must be renewed every 30 months by a certified inspector. If necessary, an approved continuous examination program (ACEP) can be used in place of this procedure.

## Transportation Security Regulations

In the post-9/11 world, a number of security initiatives have been enacted at both national and international levels that have an impact on the intermodal container industry. U.S. security initiatives include the Maritime Transportation Security Act of 2002, the

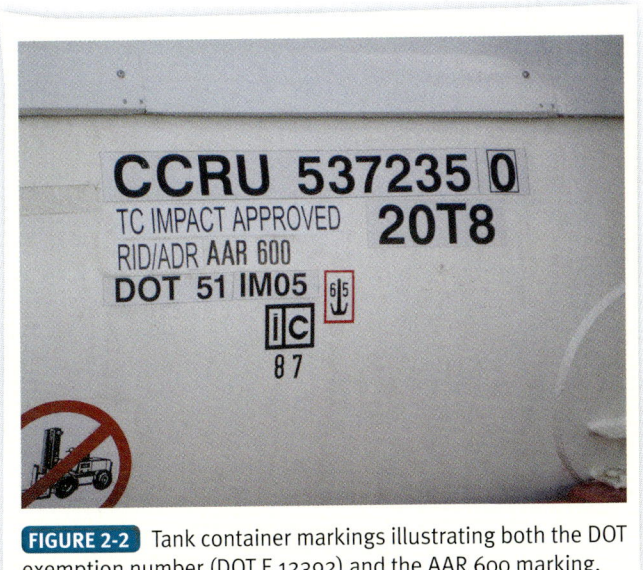

**FIGURE 2-2** Tank container markings illustrating both the DOT exemption number (DOT E 12392) and the AAR 600 marking.

Courtesy of Bill Hand.

Intelligence Reform and Prevention Act of 2006, and the Security and Accountability Act of 2006 (SAFE Port Act).

Within the United States, these initiatives have had a wide range of scope and application, including enhanced cyber security capabilities, risk mitigation requirements to enhance port resilience and recovery capabilities, and improvised explosive device (IED) and Chemical, Biological, Radiological, Nuclear, Explosive (CBRNE) prevention, protection, response, and recovery capabilities. In addition, the Transportation Worker Identification Credential (TWIC®) has been enacted by the U.S. DHS–Transportation Security Administration to ensure that individuals who may pose a threat do not gain unescorted access to secure areas of the nation's maritime transportation system. (See **FIGURE 2-4**.)

**FIGURE 2-4** Freight container undergoing screening at a marine port.

Courtesy of Bill Hand.

## Summary

Intermodal containers are designed, constructed, and handled in accordance with a number of both domestic and international codes, regulations, and consensus standards.

An integrated network of regulations, standards, and industry guidelines directly impact the intermodal container industry. Key regulations and standards that impact the intermodal industry include the Recommendations on the Transport of Dangerous Goods, which has been adopted by the U.S. Department of Transportation as law and the International Maritime Dangerous Goods (IMDG).

## References

1. Brassington, Bill. 2013. *"Safe Handling of Containers" Safety Briefing Pamphlet Series #30*. International Tank Container Organisation. Romford, Essex, United Kingdom.
2. International Tank Container Organisation. 2013 *Tank Container Fleet Survey* (June). Romford, Essex, United Kingdom.
3. Noll, Gregory G. and Michael S. Hildebrand. 2014. *Hazardous Materials: Managing the Incident*, 4th ed. Burlington, MA: Jones and Bartlett Learning, pp. 155–156.
4. United Nations, 2015 *Recommendations on the Transport of Dangerous Goods - Model Regulations*, 19th ed. New York.

a clear picture of the container and what may be inside without even touching it.

Regulations and standards require that intermodal containers convey information through markings and placards. This information centers on the engineering design and limitations of the container and the hazards of the product inside the container.

## Chapter Outline

- Key Terms
- Introduction
- Container Markings and Identification Data Plates
- Placards
- Other Types of Intermodal Container Identification Systems
- Markings Unique to Portable Tank Containers
- Summary
- References

## Key Terms

**Container Markings** Markings on both freight containers and portable tank containers that provide useful information for safe container and cargo handling as well as potential safety or environmental hazards.

**Placards** A display that conveys information about the hazards of the cargo on board a motor vehicle or rail car. This includes the hazard class and the four-digit identification number that allows emergency responders to identify the material and obtain emergency initial response guidance and information from the DOT Emergency Response Guidebook. Placards are required to be displayed on both sides and at each end of the motor vehicle or rail car.

## Introduction

To the inexperienced observer, the information displayed on an intermodal container is a mystery. But with some training and experience you can develop a clear picture of what you are dealing with. National and international standards have established a system to convey information about the container no matter what language you speak. Once you understand the system, you can develop

## Container Markings and Identification Data Plates

Markings on both freight containers and tank containers can provide useful information for safe container and cargo handling as well as potential safety or environmental hazards. There are two types of markings: 1) Those that provide information about the tank container (e.g., capacity, lifting points, general cautions); and 2) Those that provide information about the contents of the freight or portable tank container and its hazards to people and the environment. This section will provide an overview of the information that can be obtained with trained observation.

- **Data/Identification Plate**—U.S. DOT 49 CFR 178.274 requires that every portable tank container must be fitted with a corrosion-resistant metal identification plate that is permanently attached to the tank container and is usually located on the rear (discharge) end of the container. There are no data plate requirements for freight containers. There are 25 fields of information required by DOT on a data plate. All of this information is useful for shippers, carriers, and customs and enforcement personnel, but most of the information is not relevant to emergency responders. **FIGURE 3-1** shows an example of a data plate as well as the more important information that responders may want to know and why it is useful.

# INFORMATION REQUIRED ON A DATA PLATE FOR PORTABLE TANKS

Every portable tank must be fitted with a corrosion resistant metal plate permanently attached to the portable tank in a conspicuous place and readily accessible for inspection. When the plate cannot be permanently attached to the shell, the shell must be marked with at least the information required by Section VIII of the American Society of Mechanical Engineers Code (ASME).

There are U.S. DOT regulations requiring 25 fields of information required on a data plate. This information that is most useful to emergency responders includes:

**Maximum Allowable Working Pressure (MAWP)**—The MAWP provides a general idea of how the tank may be pressurized and how high the pressure inside may be. The data plate will show the MAWP "bar gauge" and can be easily converted to pounds per square inch gauge using an application on your hand held information device. For example, 1 bar gauge = 14.5 psig.

**Water Capacity at 20°C/Liter**—This information is valuable in terms of determining how much product may need to be removed from the damaged tank to the recovery tank. If spill control issues are involved the maximum potential spill size can be estimated. The water capacity information will be in liters. 1 Liter = 0.26 U.S. liquid gallons. 20°C = 68°.

**Water Capacity of Each Compartment at 20°C/Liter**—If the data plate is accessible (the tank is not overturned) this can give responders a quick idea if the portable tank has compartments without climbing on top. If only one compartment is leaking it provides information on how much product must be removed. Once the product has been recovered and the amount recovered has been determined in gallons, the amount spilled can be calculated. This information is useful to environmental agencies. The water capacity information will be in liters. 1 Liter = 0.26 U.S. liquid gallons. 20°C = 68°.

**Initial Pressure Test Date and Most Recent Retest Date**—From a safety perspective, this information helps make decisions on handling the container, especially if the pressure test date has expired. 1 bar gauge = 14.5 psig.

**Shell Material(s) and Material Standard Reference(s)**—This information is useful if drilling is being considered for product removal.

**Equivalent Thickness in Reference Steel, mm**—The shell thickness provides useful information for leak making leak control decisions. 1 Millimeter = 0.039 inches.

**Either "thermally insulated" or "vacuum insulated"**—This information may be useful for devloping  leak control tactics. If the tank is insulated there will usually be an outer aluminum jacket around the tank with the insulation placed between the outer shell wrapping and the inner tank shell. A breach of the outer jacket may appear as a penetration of the tank shell when the tank is unbreached. Insulation sometimes absorbs moisture and becomes soggy. Liquid dripping out is usually water, especially during simmer months.

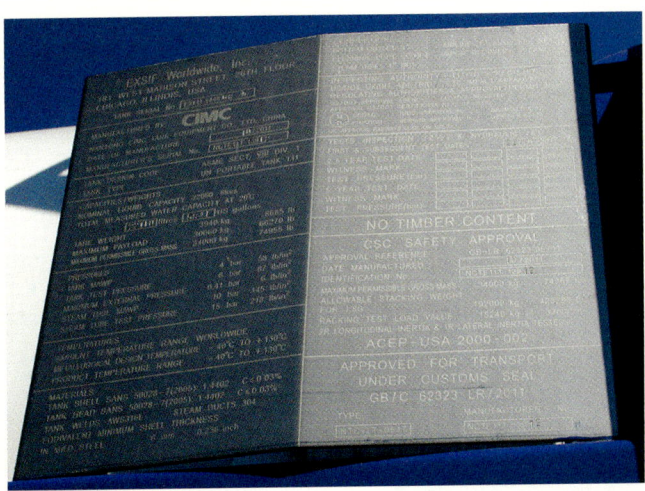

**FIGURE 3-1** Useful information for emergency responders available on a data/identification plate.

Courtesy of Bill Hand.

- **Owners Code and Number**—All freight and tank containers that are in international intermodal service must be marked with a four capital-letter code consisting of (a) a three-letter owner identification code; (b) an equipment identifier (there are several identifiers but only one for intermodal containers through—U) and (c) a serial number consisting of six Arabic numerals. The seventh digit of the container number, also known as the check digit, is calculated off the four letters and six digits of the container number. As long as the container number follows the ISO standard, the container number can be entered into a check digit calculator to determine if it is a valid number. Tank containers are registered with the Bureau International des Containers et du Transport International (B.I.C.) in France. (See **FIGURE 3-2 A**.) The initials on the Owners Code and Number indicate ownership of the container. The number is like a unique human fingerprint; it identifies the specific container. These markings are generally found on the right-hand side of the container (as you face it from either side), and on both ends. (See **FIGURE 3-2 B** and **FIGURE 3-3**.)
- **Country, Size, and Type Markings**—Portable tank containers as well as freight containers will display a country, size, and type code. *Note:* Showing the country is currently required in DOT regulations but is typically not found on older containers.

These markings will be found in conjunction with the Owners Code and Number, and are located on the right-hand side of the container (as you face it from either side), and on both ends. On freight containers they will be found at the end of the container. (See **FIGURES 3-4** and **3-5**.)

ISO Standard 3166 is the International Standard for country codes and their subdivisions. The system uses two or three letters to identify a country rather than spell out the entire name. This saves time and reduces errors because the letters are readable by machine. For example, postal services throughout the world exchange international mail in containers identified with their respective country code. These codes are used in banking, passports, etc.

In the intermodal container business, these standardized country codes indicate the country of registry for the container. However, responders should note the following disclaimers. First, the country of registration can vary within a company's container fleet. Second, the country code may not always appear on the container because it may not be required. Examples of some of the most common country codes are listed in **TABLE 3-1**.

The ISO size/type code indicates the container's size, characteristics, and other specific features of the container. The first two digits indicate the container's length and height; specific sizes are assigned specific numbers. The second two characters represent the type of freight or tank container. See **SCAN SHEET 3-A** for examples of these ISO Type Codes.

**FIGURE 3-2** (A) The Check Digit is located at the end of the numerical sequence. (B) Owners Code and Number (EXFU) and the serial number (056064). The check digit is the box (3) on the right side of the container.

Courtesy of Hildebrand and Noll Associates, Inc.

**FIGURE 3-3** Owners Code and Number (ICHU) and the serial number (000165) check digit in the box (0) on the end of the container.

Courtesy of Hildebrand and Noll Associates, Inc.

## Placards

The requisite skills and knowledge for recognizing and identifying hazardous materials through the use of placards is a basic first responder competency. Placards are used to identify the hazard class or division of a hazardous material. The hazard class or division number must be displayed in the lower corner of a placard and is required for the nine primary hazard classes and subsidiary hazard classes and divisions. The hazard class of dangerous goods is indicated either by its class (or division) number or name. It is not uncommon to find placards written in a foreign language. (See **FIGURE 3-6**.) More information on specialized intermodal container placards will be covered in Chapters 4 and 5.

## Other Types of Intermodal Container Identification Systems

Intermodal freight and tank containers transporting hazardous materials and dangerous goods can be identified through the use of several different hazard marking systems, including the ADR/RID marking system and the HAZCHEM marking system. Additional information on each is found below.

### ■ ADR/RID Hazard Marking System

The European Agreement Concerning Transport of Dangerous Goods by Truck (ADR) and European Agreement Concerning Transport of Dangerous Goods

ISO Standard 6346 standardizes the Container Type Group by various types of containers and their serial number, owner, country code, and size of any given shipping container.

This standard is maintained by the International Container Bureau and covers the serial number, owner, country code, and size of any given shipping container. There are over 100 different ISO shipping container size codes. See Scan Sheet 3-A for some examples.

**FIGURE 3-4** Country Code (GB) United kingdom ISO Size/Type Code (2276).

Courtesy of Hildebrand and Noll Associates, Inc.

| Table 3-1 | Common Country Codes |
|---|---|
| **ISO Code** | **Country** |
| BM | Bermuda |
| DE | Germany |
| DK | Denmark |
| FR | France |
| KP | Korea |
| HK | Hong Kong |
| IT | Italy |
| JP | Japan |
| LR | Liberia |
| NL | Netherlands |
| NZ | New Zealand |
| PA | Panama |
| RU | Russian Federation |
| PH | Philippines |
| GB | United Kingdom |
| SG | Singapore |
| SE | Sweden |
| CH | Switzerland |
| US | United States |

by Rail (RID) are used within Europe and some South American countries for the movement of portable tank containers. These regulations are in line with the IMDG Code requirements, and are used to assist emergency responders in safely identifying the contents and hazards of a hazardous materials transportation container

**FIGURE 3-5** Country Code missing/not required.

Courtesy of Hildebrand and Noll Associates, Inc.

**FIGURE 3-6** Examples of foreign language placards. It is very common to see these placards at U.S. and international ports around the world.

Courtesy of Hildebrand and Noll Associates, Inc.

or vehicle. Intermodal tanks and containers shipped into the United States and North America may contain these markings. The ADR/RID Marking System is outlined in SCAN SHEET 3-B .

The ADR/RID Marking System consists of two orange panels with black printing. The upper box contains a Hazard Identification Number, while the lower box contains the four-digit United Nations identification number. This four-digit number can easily be researched in the Emergency Response Guidebook. (See FIGURE 3-7 .)

The Hazard Identification Number (upper panel) consists of two or three digits and sometimes the letter X.

The first digit indicates the primary hazard. In general, the digits indicate the following hazards:

1. Explosive risk
2. Emission of gas due to pressure or chemical reaction
3. Flammability of liquids (vapors) and gases or self-heating liquid
4. Flammability of solids or self-heating solid
5. Oxidizing (fire-intensifying) effect
6. Toxicity or risk of infection
7. Radioactivity
8. Corrosivity
9. Risk of spontaneous violent reaction

Note the following additional information:

- The risk of spontaneous violent reaction within the meaning of digit 9 includes the possibility, due to the nature of a substance, of a risk of explosion, disintegration, and polymerization reaction followed by the release of considerable heat or flammable and/or toxic gases.

- Doubling of a digit indicates an intensification of that particular hazard. For example 33 indicates that the product is extremely flammable, 66 indicates higher toxicity, and 88 indicates the product is highly corrosive.
- Where the hazard associated with a substance can be adequately indicated by a single digit, the digit is followed by a zero (e.g., 30, 40, 50).
- A hazard identification number prefixed by the letter "X" indicates that the substance will react dangerously with water. For example, X88 indicates that the product is highly corrosive and will also react with water.

## ■ HAZCHEM Marking System

HAZCHEM is a hazard marking system used in Australia, Malaysia, New Zealand, and the United Kingdom for vehicles transporting hazardous substances. The top-left section of the plate gives the Emergency Action Code (EAC) telling the emergency responders what actions to take during an emergency response. The middle-left section gives the UN Identification Number describing the product. The lower-left section gives the telephone number that should be called if special advice is needed. The placard at top-right indicates the hazard class of the product. (See FIGURES 3-8 and 3-9 .)

## Markings Unique to Portable Tank Containers

In addition to the markings required on all intermodal containers, such as the Owners Code and Number and the Country, Size and Type Markings, tank containers

**FIGURE 3-7** The ADR/RID Marking System.
Courtesy of Hildebrand and Noll Associates, Inc.

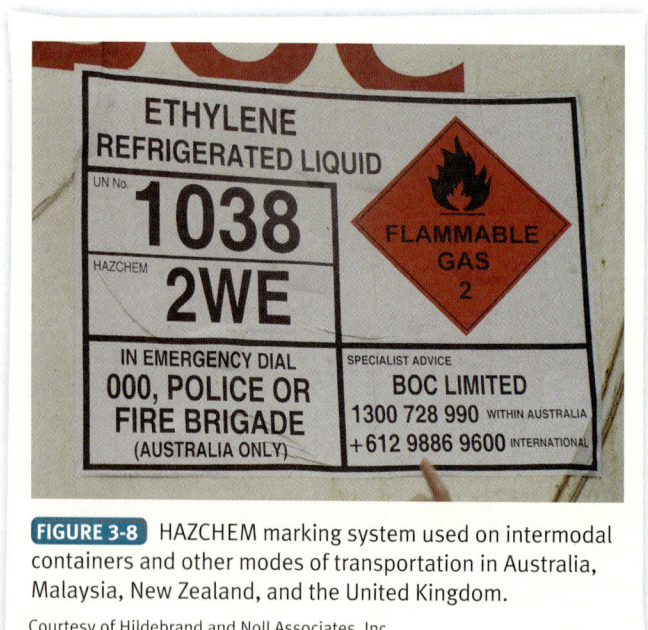

**FIGURE 3-8** HAZCHEM marking system used on intermodal containers and other modes of transportation in Australia, Malaysia, New Zealand, and the United Kingdom.
Courtesy of Hildebrand and Noll Associates, Inc.

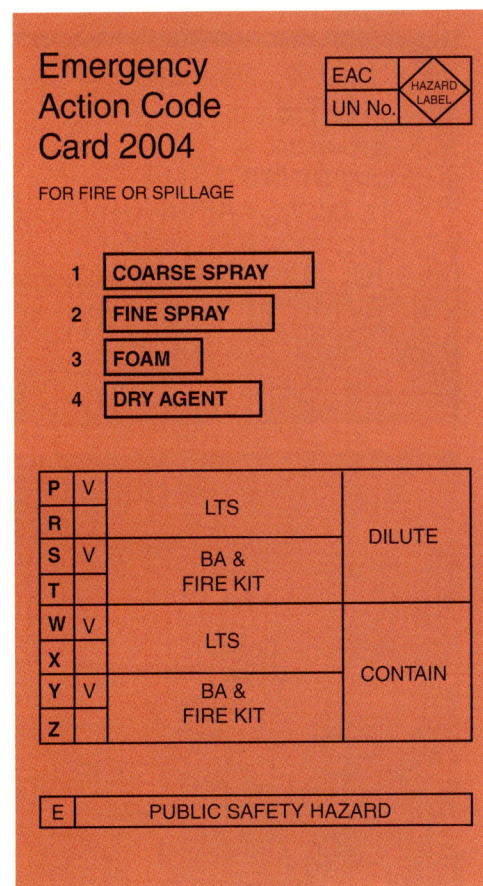

**FIGURE 3-9** HAZCHEM Emergency Action Codes (EAC) pocket card.

Courtesy of The National Chemical Emergency Centre (NCEC).

will have additional markings that pertain to the tank's design and construction features.

- **Specification Markings**—The specification marking indicates the standards to which a portable tank was built. Tank containers must meet UN portable tank design, construction, and safety standards. The specification mark will be on both sides of the tank, generally near the tank's Owners Code and Number. (See **FIGURE 3-10**.)

Examples of specification markings are:

- UN Portable Tank (with or without the T code, since T code is not required but is allowed with the specification marking)
- IM-101 (25.4 psig–100 psig)
- IM-102 (14.5 psig–25.4 psig)
- Spec. 51 (100 psig–500 psig)

You may also see the IMO Type 1, 2, 5, and 7 specifications. All are acceptable internationally as long as they are the same as the UN recommendations.

- **Weight Markings**—The maximum gross and tare weights measured in kilograms (kg) and pounds (lbs). (See **FIGURE 3-11**.)

- **Overhead Hazard Marking**—A warning of overhead electrical danger or obstruction. (See **FIGURE 3-12**.)
- **DOT Special Permit**—Marking exemptions are sometimes authorized from DOT regulations. In these cases, the outside of each package/container must be plainly and durably marked "DOT-SP" followed by the exemption number assigned (e.g., DOT SP 12392). On intermodal tanks, these markings must be in 2-inch letters. (See **FIGURE 3-13**.)
- **Minimum Tank Design Pressure**—The minimum tank design pressure is 35 psi (2.4 bar), but in no case less than the vapor pressure of the commodity at 115°F (46.1°C).

## ■ AAR 600 Marking

Tanks meeting these requirements will display the "AAR 600" marking in 2-inch letters on both sides near the tank's reporting marks and number. The "AAR 600" marking indicates tanks that can be used for regulated materials, while the "AAR-600NR" marking

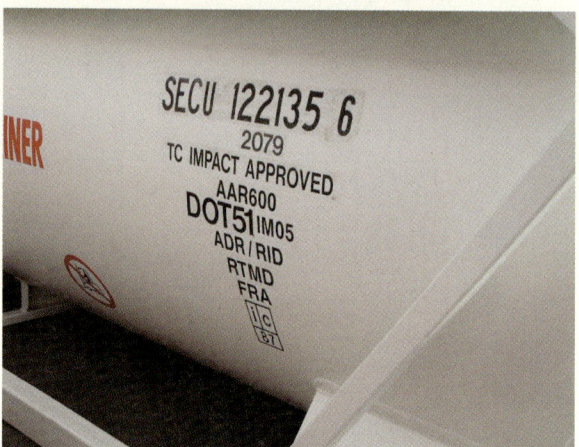

**FIGURE 3-10** Examples of Specification Markings on tank containers.

Courtesy of Hildebrand and Noll Associates, Inc.

**FIGURE 3-11** (A) Weight Markings located on the rear of a freight container. (B) Weight and Capacity Markings on the end of a tank container.

Courtesy of Hildebrand and Noll Associates, Inc.

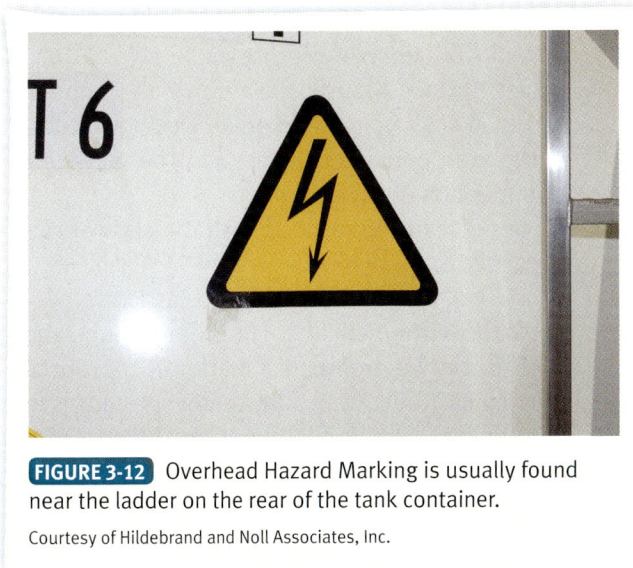

**FIGURE 3-12** Overhead Hazard Marking is usually found near the ladder on the rear of the tank container.

Courtesy of Hildebrand and Noll Associates, Inc.

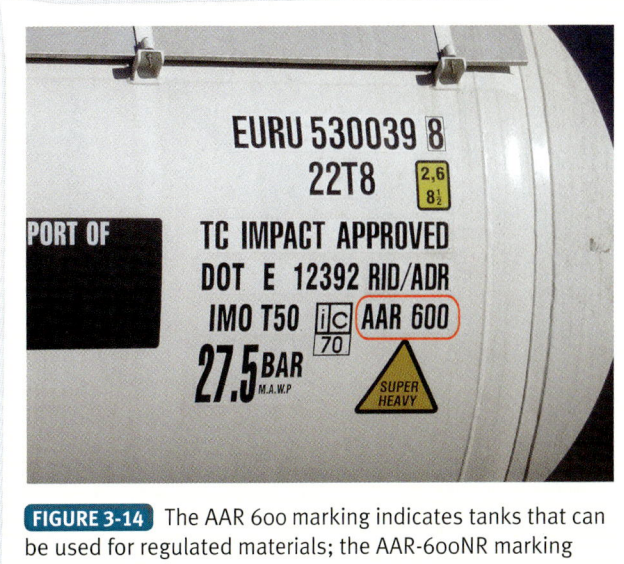

**FIGURE 3-14** The AAR 600 marking indicates tanks that can be used for regulated materials; the AAR-600NR marking indicates tanks that cannot be used for regulated materials.

Courtesy of Hildebrand and Noll Associates, Inc.

indicates tanks that cannot be used for regulated materials. (See **FIGURE 3-14**.)

- **TC Impact Marking**—Indicates the approval of Transport Canada (TC), which requires a longitudinal impact test. More information can be found in Appendix E, National Standards of Canada CAN/CGSB-43.147-2002. (See **FIGURE 3-15**.)
- **European Railway System Approval Decal UIC**—A certification that the tank container has been approved for transport on the European Railway System. The country of registration is displayed on the bottom half of the marking. (See **FIGURE 3-16**.)

- **The Convention for Safe Containers (CSC)**—Requires all tank containers that are international intermodal service carry a Safety Approval Plate with at least the following:
  - An approval reference
  - The date of manufacture
  - A manufacturer's identification number
  - Maximum operation gross weight
  - Allowable stacking weight
  - Transverse racking test load value
  - The next examination date or Approved Continuous Examination Program reference number (See **FIGURE 3-17**.)

**FIGURE 3-13** DOT Special Permit number is located on the side of the tank container below the owners Code and Size/Type Code. On this container the number is DOT E SP 12392.

Courtesy of Hildebrand and Noll Associates, Inc.

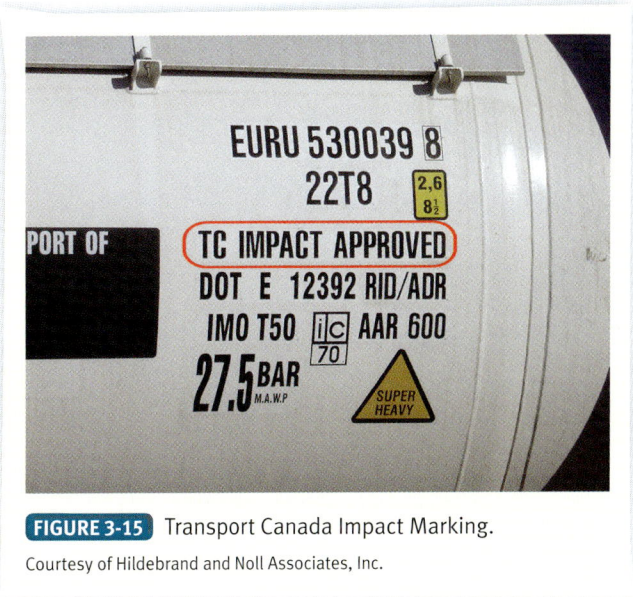

**FIGURE 3-15** Transport Canada Impact Marking.

Courtesy of Hildebrand and Noll Associates, Inc.

**FIGURE 3-16** This Railway Approval Decal UIC is a certification that the tank container has been approved for transport on the European Railway System.

Courtesy of Hildebrand and Noll Associates, Inc.

- **Customs Convention for Containers**—Requires all tank containers that are international intermodal service to carry:
  - A means to identify the owner and their address (This may be an address plate, decal or the three letter owner identification code registered with B.I.C.)
  - A customs approval number

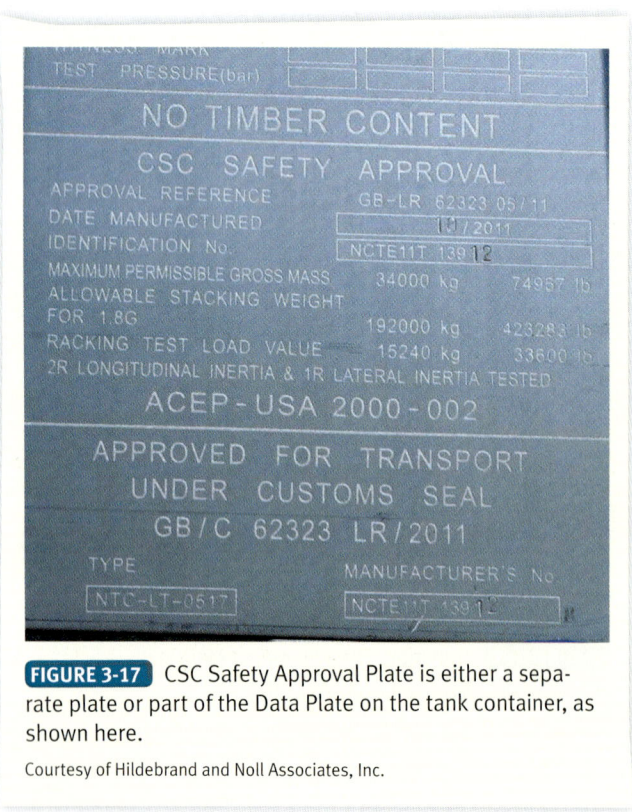

**FIGURE 3-17** CSC Safety Approval Plate is either a separate plate or part of the Data Plate on the tank container, as shown here.

Courtesy of Hildebrand and Noll Associates, Inc.

- **IMDG Code Markings**—Under the IMDG Code, all tank containers must display:
  - The Proper Shipping Name of the contents, which shall be marked on at least both sides of the tank container. It may also be marked on both ends of the container.
  - Any placards associated with the class of cargo carried.
  - The UN number for the cargo carried.
  - An Environmentally Hazardous Substances (Aquatic Environment) mark, if appropriate. For maritime transport these substances are known as Marine Pollutants and subject to the provisions of Annex III of MARPOL 73/78. (See **FIGURE 3-18**.)
  - Tank containers containing a substance that if transported in a liquid state at a temperature equal to or exceeding 212°F (100°C) or in a solid state at a temperature equal to or exceeding 464°F (240°C) shall bear on each side and on each end the mark shown in **FIGURE 3-19**. In addition to the elevated temperature mark, the maximum temperature of the substance expected to be reached during transport shall be durably marked on both sides of the tank container, immediately adjacent to the elevated temperature mark, in characters at least 4 in. (100 mm) high. (See Figure 3-19.)

Under UIC, all tank containers that have a maximum gross mass equal to or greater than 75,000 lbs (34,000 kg) must carry a super heavy decal. (See **FIGURE 3-20**.)

## Optional Markings

- **Height Markings**—A mark that details the height of a container. The marks consist of sets of black figures on a yellow background surrounded by a black border. The upper set of figures gives the height in meters to one decimal place (0.1 m), which will not be less than the actual height. The lower set of figures gives the height in feet to the nearest 1/4 ft., which is not less than the actual height. It is displayed on the right sides and right ends of the container. (See **FIGURE 3-21**.)
- **Calibration Chart (strapping chart)**—A chart to provide guidance on the volume of cargo based on a dipstick measurement. This is usually located inside the spill box near the manhole on top of the nonpressure tank container.
- **Do Not Walk Decal**—Insulated containers may have a decal similar to that shown in **FIGURE 3-22**.

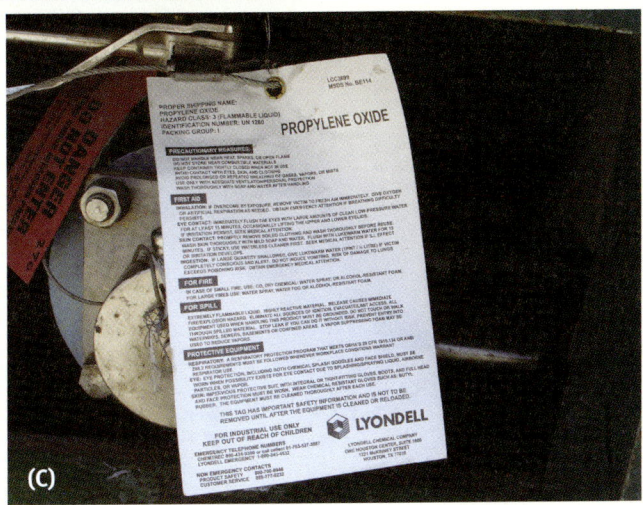

**FIGURE 3-18** A. Proper Shipping Name. B. Environmentally Hazardous Substance Display. C. Product information tag attached to discharge valve.

Courtesy of Hildebrand and Noll Associates, Inc.

**FIGURE 3-19** Elevated Temperature Mark and Maximum Temperature.

Courtesy of Hildebrand and Noll Associates, Inc.

**FIGURE 3-20** Super Heavy Decal.

Courtesy of Hildebrand and Noll Associates, Inc.

**FIGURE 3-21** Height Markings.

Courtesy of Hildebrand and Noll Associates, Inc.

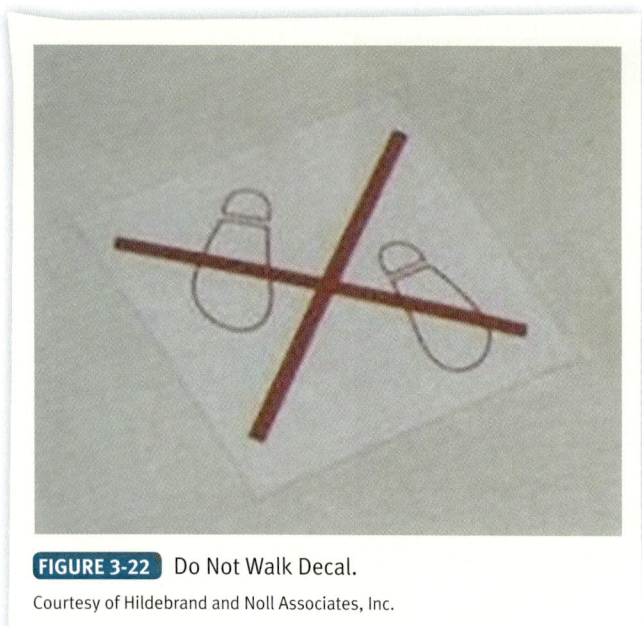

**FIGURE 3-22** Do Not Walk Decal.

Courtesy of Hildebrand and Noll Associates, Inc.

**FIGURE 3-23** Do Not Use Fork Truck Decal.

Courtesy of Hildebrand and Noll Associates, Inc.

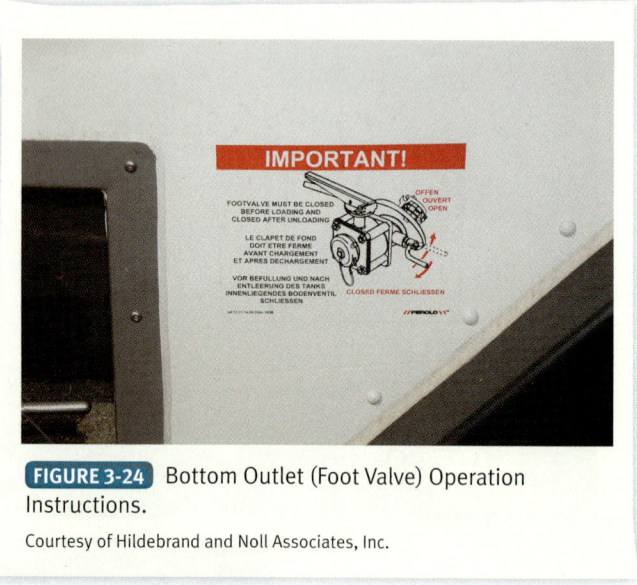

**FIGURE 3-24** Bottom Outlet (Foot Valve) Operation Instructions.

Courtesy of Hildebrand and Noll Associates, Inc.

**FIGURE 3-25** Food Grade Cargo Decal.

Courtesy of Hildebrand and Noll Associates, Inc.

affixed to ensure that the insulation cladding is not stepped on. Stepping on it could be a slip hazard and damage the outer jacket and the insulation material.

- **Do Not Use Fork Truck Decal**—Containers that may be damaged by fork lift vehicles are marked with a decal shown in **FIGURE 3-23**.
- **Valve Operation Instructions**—This decal shows how to operate the bottom outlet (foot valve) and is required by RID/ADR. (See **FIGURE 3-24**.)
- **Food Grade Cargo Decal**—Containers transporting food grade cargo are required to post a decal with rectangle and the name of the food grade cargo being carried. (See **FIGURE 3-25**.)

© Photos.com/Getty

## Container Size Code

ISO Standard 6346 also standardizes the Container Type by Group and associates that information with standardized container sizes and a general description.

There are over 50 different container types and numerous container sizes and descriptions. The codes are stenciled on the outside of the container.

Courtesy of Hildebrand and Noll Associates, Inc.

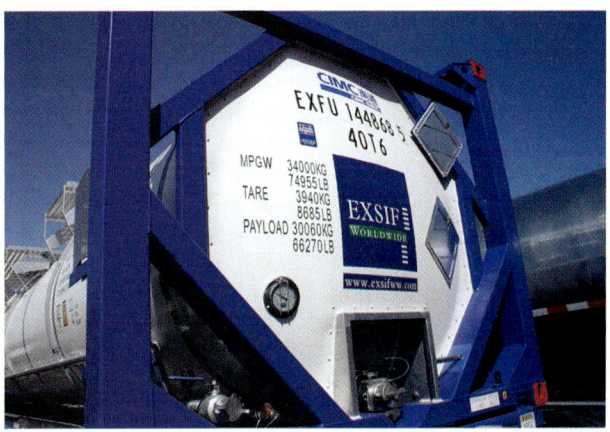

Courtesy of Hildebrand and Noll Associates, Inc.

| Container Type Group | | Size Type |
| --- | --- | --- |
| **Code** | **Description** | **Code** |
| 20GP | General Purpose (Freight) Container | 20G0 |
| 20HR | Insulated container | 20G1 |
| 20PF | Flat (Fixed Ends) | 20P1 |
| 20D | Tank Container | 20T3 |
| 22RT | Refrigerated Container | 22R1 |
| 22UT | Open Top Container | 22U1 |
| 45BK | Bulk Carrier | 4583 |
| 48TG | Tank for Gas | 48T8 |

Note: For each Container Type Group there may be several different Size Types. We only show one Size Type above for each Group.

NOTE: Doubling of a digit indicates an intensification of that particular hazard (e.g., 33 would indicate the flammable risk is higher). Where the hazard is associated with a substance and can be adequately indicated by a single figure, the figure will be followed by a zero (e.g., 30).

The second and third digits indicate the secondary or tertiary hazards.

In general, the numerals indicate the following hazards:

| | |
|---|---|
| 1 | Explosive risk |
| 2 | Emission of gas due to pressure or chemical reaction |
| 3 | Flammability of liquids (vapors) and gases or self-heating liquid |
| 4 | Flammability of solids or self-heating solid |
| 5 | Oxidizing (fire-intensifying) effect |
| 6 | Toxicity or risk of infection |
| 7 | Radioactivity |
| 8 | Corrosivity |
| 9 | Risk of spontaneous violent reaction |

The hazard identification numbers listed below have special meaning and provide more detailed information.

| | |
|---|---|
| 20 | Asphyxiant gas or gas with no subsidiary risk |
| 22 | Refrigerated liquefied gas, asphyxiant |
| 223 | Refrigerated liquefied gas, flammable |
| 225 | Refrigerated liquefied gas, oxidizing (fire-intensifying) |
| 23 | Flammable gas |
| 236 | Flammable gas, toxic |
| 239 | Flammable gases, which can spontaneously lead to violent reaction |
| 25 | Oxidizing (fire-intensifying) gas |
| 26 | Toxic gas |
| 263 | Toxic flammable |
| 265 | Toxic gas, oxidizing (fire-intensifying) |
| 268 | Toxic gas, corrosive |
| 286 | Corrosive gas, toxic |
| 28 | Gas, corrosive |

| | |
|---|---|
| 30 | Flammable liquid (flash-point between 23°C and 60°C, inclusive), or flammable liquid or solid in the molten state with a flash point above 60°C, heated to a temperature equal to or above its flash point, or self-heating liquid |
| 323 | Flammable liquid which reacts with water, emitting flammable gases |
| X323 | Flammable liquid which reacts dangerously with water, emitting flammable gases |
| 33 | Highly flammable liquid (flash point below 23°C) |

| | |
|---|---|
| 333 | Pyrophoric liquid |
| X333 | Pyrophoric liquid which reacts dangerously with water* |
| 336 | Highly flammable liquid, toxic |
| 338 | Highly flammable liquid, corrosive |
| X338 | Highly flammable liquid, corrosive which reacts dangerously with water* |
| 339 | Highly flammable liquids which can spontaneously lead to violent reaction |
| 36 | Flammable liquid (flash-point between 23°C and 60°C, inclusive), slightly toxic, or self-heating liquid, toxic |
| 362 | Flammable liquid, toxic, which reacts with water, emitting flammable gases |
| X362 | Flammable liquid, toxic, which reacts dangerously with water, emitting flammable gases |
| 38 | Flammable liquid (flash-point between 23°C and 60°C, inclusive), slightly corrosive or self-heating liquid, corrosive |
| 382 | Flammable liquid, corrosive, which reacts with water, emitting flammable gases |
| X382 | Flammable liquid, corrosive, which reacts dangerously with water, emitting flammable gases |
| 39 | Flammable liquids which can spontaneously lead to violent reaction |

| | |
|---|---|
| 40 | Flammable or self-heating solid |
| 423 | Solid which reacts with water, emitting flammable gases |
| X423 | Solid which reacts dangerously with water, emitting flammable gases, or flammable solid which reacts dangerously with water, emitting flammable gases, or self-heating solid which reacts dangerously with water, emitting flammable gases |
| 43 | Spontaneously flammable (pyrophoric) solid |
| X432 | Spontaneously flammable (pyrophoric) solid which reacts dangerously with water, emitting flammable gases |
| 44 | Flammable solid in the molten state at an elevated temperature |
| 446 | Flammable solid, toxic, in the molten state at an elevated temperature |
| 46 | Flammable or self-heating solid, toxic |
| 462 | Toxic solid which reacts with water, emitting flammable gases |
| X462 | Solid which reacts dangerously with water, emitting toxic gases |
| 48 | Flammable or self-heating solid, corrosive |
| 482 | Corrosive solid which reacts with water, emitting flammable gases |
| X482 | Solid which reacts dangerously with water, emitting corrosive gases |

| | |
|---|---|
| 50 | Oxidizing (fire-intensifying) substance |
| 539 | Flammable organic peroxide |
| 55 | Strongly oxidizing (fire-intensifying) substance |
| 556 | Strongly oxidizing (fire-intensifying) substance, toxic |
| 558 | Strongly oxidizing (fire-intensifying) substance, corrosive |
| 559 | Strongly oxidizing (fire-intensifying) substance which can spontaneously lead to violent reaction |
| 56 | Oxidizing substance (fire-intensifying), toxic |
| 568 | Oxidizing substance (fire-intensifying), toxic, corrosive |
| 58 | Oxidizing substance (fire-intensifying), corrosive |
| 59 | Oxidizing substance (fire-intensifying) which can spontaneously lead to violent reaction |

| | |
|---|---|
| 60 | Toxic or slightly toxic substance |
| 606 | Infectious substance |
| 623 | Toxic liquid, which reacts with water, emitting flammable gases |
| 63 | Toxic substance, flammable (flash-point between 23°C and 60°C, inclusive) |

| | |
|---|---|
| 638 | Toxic substance, flammable (flash-point between 23°C and 60°C, inclusive), corrosive |
| 639 | Toxic substance, flammable (flash-point not above 60°C), which can spontaneously lead to violent reaction |
| 64 | Toxic solid, flammable or self-heating |
| 642 | Toxic solid which reacts with water, emitting flammable gases |
| 65 | Toxic substance, oxidizing (fire-intensifying) |
| 66 | Highly toxic substance |
| 663 | Highly toxic substance, flammable (flash point not above 60°C) |
| 664 | Highly toxic solid, flammable or self-heating |
| 665 | Highly toxic substance, oxidizing (fire-intensifying) |
| 668 | Highly toxic substance, corrosive |
| X668 | Highly toxic substance, corrosive, which reacts dangerously with water |
| 669 | Highly toxic substance which can spontaneously lead to violent reaction |
| 68 | Toxic substance, corrosive |
| 69 | Toxic or slightly toxic substance which can spontaneously lead to violent reaction |

| | |
|---|---|
| 70 | Radioactive material |
| 78 | Radioactive material, corrosive |

| | |
|---|---|
| 80 | Corrosive or slightly corrosive substance |
| X80 | Corrosive or slightly corrosive substance which reacts dangerously with water |
| 823 | Corrosive liquid which reacts with water, emitting flammable gases |
| 83 | Corrosive or slightly corrosive substance, flammable (flash point between 23°C and 60°C, inclusive) |
| X83 | Corrosive or slightly corrosive substance, flammable (flash point between 21°C and 55°C), which reacts dangerously with water* |
| 839 | Corrosive or slightly corrosive substance, flammable (flash point between 23°C and 60°C, inclusive), which reacts dangerously with water* |
| X839 | Corrosive or slightly corrosive substance, flammable (flash point between 23°C and 60°C, inclusive), which can spontaneously lead to violent reaction and which reacts dangerously with water* |
| 84 | Corrosive solid, flammable or self-heating |
| 842 | Corrosive solid which reacts with water, emitting flammable gases |
| 85 | Corrosive or slightly corrosive substance, oxidizing (fire-intensifying) |
| 856 | Corrosive or slightly corrosive substance, oxidizing (fire-intensifying) and toxic |
| 86 | Corrosive or slightly corrosive substance, toxic |
| 88 | Highly corrosive substance |
| X88 | Highly corrosive substance which reacts dangerously with water |
| 883 | Highly corrosive substance, flammable (flash point between 23°C and 60°C, inclusive) |
| 884 | Highly corrosive solid, flammable or self-heating |
| 885 | Highly corrosive substance, oxidizing (fire-intensifying) |
| 886 | Highly corrosive substance, toxic |
| X886 | Highly corrosive substance, toxic, which reacts dangerously with water |
| 89 | Corrosive or slightly corrosive substance which can spontaneously lead to violent reaction |

| | |
|---|---|
| 90 | Environmentally hazardous substance; miscellaneous dangerous substances |
| 99 | Miscellaneous dangerous substance carried at an elevated temperature |

*Water not to be used except by approval of the competent authority.

## Summary

Intermodal container markings and placards are critical elements in identifying the hazardous materials involved and evaluating the hazards and risks of the incident scenario.

Containers have many different types of required and optional marking systems that can provide useful information for safe container and cargo handling and potential safety or environmental hazards that may be present to the cargo handler and emergency responders.

The Owner Code and Number is one of the most important markings on any container. This number identifies the owner of the container, which can lead to establishing communications with product and container specialists who can provide technical advice to responders. This marking is registered with the Bureau International des Containers et du Transport International (B.I.C.) in France and provides important information about the shipper and contents of the container.

Other markings, such as the size/type code, provide information as to the size and type of container for both tank and freight containers. Placards and product information decals provide shippers and emergency responders important information as to the hazards of the lading. Some optional decals provide information about container handling.

## References

1. Brassington, Bill. 2013. *"Safe Handling of Containers" Safety Briefing Pamphlet Series #30*. International Tank Container Organisation. Romford, Essex, United Kingdom. pp. 27–29.
2. U.S. Department of Transportation. 2016. *Emergency Response Guidebook*, Washington, DC.
3. ISO 6346, International Standards Organization.
4. ISO 3162, International Standards Organization.
5. The Tank Container Association. 2000. *Tank Container Repair Guidelines and Definitions*. pp. 51–83.

# CHAPTER 4

# Intermodal Freight Containers

## Chapter Outline

- Key Terms
- Introduction
- Freight Container Construction Features
- Nonbulk Packaging
- Bulk Packaging That May Be Placed on/in Transport Vehicles
- Summary
- References

## Key Terms

**Dry Bulk Cargo** Cargo that may be loose, granular, free-flowing, or solid, and is shipped in bulk rather than in package form. Dry bulk cargo is usually handled by specialized mechanical handling equipment at specially designed dry bulk terminals.

**Freight Container** An intermodal container for transporting packages in unit form with a volume of 64 cubic ft. (1.8 cubic meters) or more, intended primarily for containment of packages in unit form during transportation. See the glossary for a more detailed definition.

## Introduction

The most common intermodal container in circulation throughout the world is the intermodal freight container. Within the transportation industry, intermodal freight containers may also be known as a dry van, or a box container.

There are an estimated 17 million intermodal freight containers in the world today, with different types of application and use. Freight containers are the "heavy lifters" of the intermodal cargo industry, and are used for transporting a wide range of both hazardous and nonhazardous materials. From an emergency response and law enforcement perspective, they share many of the same response issues and concerns as regularly encountered with vans, box trucks, and tractor-trailers in the highway system.

Freight containers easily move between the marine, rail, and highway transportation systems. Depending upon the mode of transportation, they can be found locked in place on the transport vehicle (highway, rail, and marine), stacked for marine and rail transportation, or stacked at ports and fixed facilities for short- or long-term storage. (See **FIGURE 4-1 A-D**.)

## Freight Container Construction Features

The key construction features of intermodal freight containers include the following:

- **Container Dimensions**—Freight containers are commonly found in two lengths—20 ft. and 40 ft., although containers as long as 45 ft., 48 ft., and 53 ft. may also be found. In some countries, containers may also include 10 ft. and 30 ft. lengths. All ISO containers are 8 ft. wide. In addition to the standardized length and width, ISO also specifies standardized heights.

  The ISO standardized sizes are summarized in **TABLE 4-1** below.

  The dimension of an 8 ft. × 8 ft. × 20 ft. intermodal freight container is used as the basic measuring stick in the intermodal cargo business. This standard container is referred to as a "TEU," meaning that it is a Twenty-foot Equivalent Unit. Therefore, an 8 ft. × 8 ft. × 40 ft. intermodal freight container is equal to two TEUs. As an example, a marine vessel capable of handling four hundred 40-foot containers is rated at 800 TEUs. The term

**FIGURE 4-1** Intermodal freight containers are found in marine, highway, and rail modes of transportation.
Courtesy of Hildebrand and Noll Associates, Inc.

| Table 4-1 | ISO Standard Freight Container Dimensions | |
|---|---|---|
| LENGTH | WIDTH | HEIGHT |
| 20 ft. | 8 ft. | 8 ft. |
| 20 ft. | 8 ft. | 8 ft., 6 in. |
| 40 ft. | 8 ft. | 8 ft., 6 in. |
| 40 ft. | 8 ft. | 9 ft., 6 in. |
| 40 ft. | 8 ft. | 8 ft., 6 in. |
| | | 4 ft., 3 in.* |
| | | *Referred to as half-heights |

"FEU" may also be used, which refers to Forty-foot Equivalent Units. Note that while 20- and 40-foot freight containers are the most common, there are also other units that range from 45 to 48 to 53 ft. long. (See **FIGURE 4-2**.)

- **Freight Container Doors**—Door designs on intermodal freight containers vary widely, as do the number of locking assemblies. Doors may have two, three, or four latch bar assemblies. (See **FIGURE 4-3 A and B**.) Although fairly common in Europe, side doors are typically not found on containers in North America. (See **FIGURE 4-3 C**.)

Since 1990, almost all standard freight containers have been manufactured with plywood floors. Most modern 40 ft. freight containers have a "tunnel" floor design, which is a raised portion of the floor at the front of the container that is designed to go on a gooseneck trailer chassis. This lowers the overall height of the container chassis combination for road travel.

## ■ Materials of Construction

There are two general types of freight container construction: (1) corrugated-side containers, or (2) smooth-side containers.

- **Corrugated-Side Freight Containers**—A typical standard 40-foot freight container is manufactured from weathering steel, which is corrosion-resistant steel that is constructed using a number

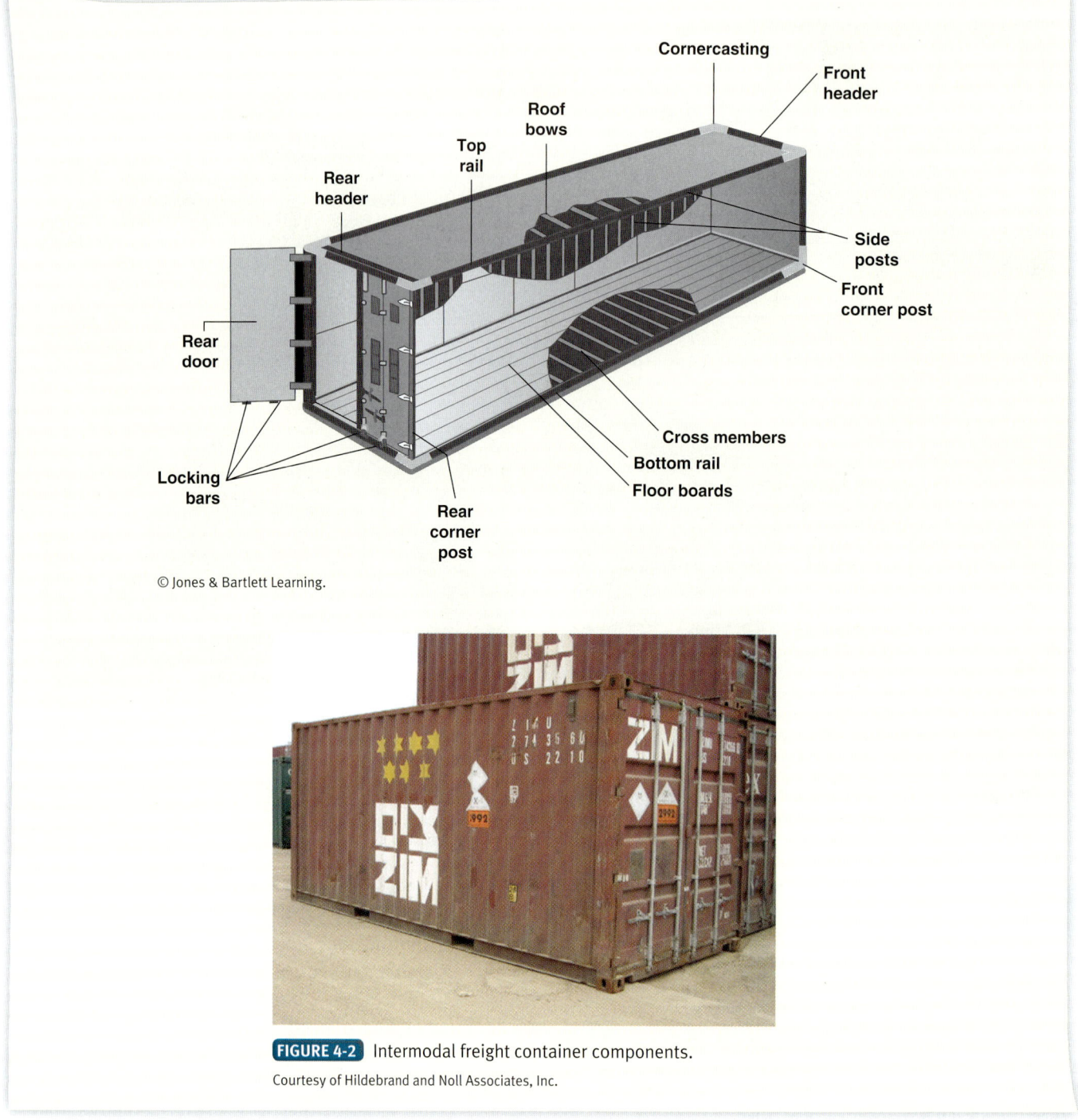

© Jones & Bartlett Learning.

**FIGURE 4-2** Intermodal freight container components.

Courtesy of Hildebrand and Noll Associates, Inc.

of welded corrugated panels. These external corrugations may be either a flat, square design or beveled with the corrugations angled or nearly rounded. Most of these containers are vented, with the vents appearing as small rectangular panels found near the corners of the container sides, or as the filler between the corrugations. (See **FIGURE 4-4 A** .) Some specialized open-top freight containers may be found with a canvas cover. (See **FIGURE 4-4 B** .)

- **Smooth-Side Freight Containers**—These are generally made of aluminum or fiberglass reinforced plywood (FRP). Aluminum smooth-side containers are constructed of a number of panels with interior posts riveted to them, while the FRP containers have no visible rivets or corrugations. Most of these containers are unvented. Vented smooth-side containers can be identified by a series of small holes along the top or bottom of the side panels. (See **FIGURE 4-4 C** .)

© Jones & Bartlett Learning.

Courtesy of Hildebrand and Noll Associates, Inc.

© topae/Shutterstock.

**FIGURE 4-3** Intermodal freight doors are normally located at the end of the container, but may sometimes be found on the side on special containers.

- **External-Post Containers**—External-post containers have a number of posts riveted to the outside of the container. These are typically constructed of steel or aluminum. Vented external-post containers can be identified by the small rectangular panels found usually near the corners of the container sides. (See **FIGURE 4-4 D** .)

## ■ Corner Castings

In intermodal container yards, most shipping containers are handled by top spreaders using what's called corner fittings or corner castings. Supporting frames for all intermodal containers, freight containers, and portable tanks are built with corner castings fittings. (On containers longer than 40 ft. the corner castings may not be 90 degrees on the corner but on the side.) They are used to secure the container in rail and highway transit. (See **FIGURE 4-5** .) On container ships, freight containers may be secured using corner castings and a lashing system. Corner castings are also used to lift freight containers with standard container handling equipment.

**FIGURE 4-4** Corrugated, smooth-side, and external-post intermodal freight containers.
Courtesy of Hildebrand and Noll Associates, Inc.

**FIGURE 4-5** Corner castings.
Courtesy of Hildebrand and Noll Associates, Inc.

Corner castings must conform to ISO Standard 1161—*Specifications of Corner Fittings for Series Freight Containers*. Cast iron corner castings are prohibited.

### ■ Container Floors

Shipping container floors are typically made of planking or plywood, which is strong and resilient, does not dent easily, and can be replaced during repairs or if contaminated. The floors provide a strong friction surface, which is important for securing cargo. Most wooden container floors are sprayed for insects because when lumber is used, it must comply with quarantine regulations.

### ■ Cargo Capacities and Weights

A standard 40-foot FEU can carry about 2,400 cubic ft. (68 m³) of cargo. The gross weight of a typical corrugated metal container is approximately 67,200 pounds (30,481 kg) with an average tare weight of 8,630 pounds (3,914 kg). These numbers will vary depending on the manufacturer or modifications.

The *gross weight* of a container is the total weight of the container itself plus the maximum weight of the cargo that can be carried inside. *Tare* Weight is the weight of the container empty with no cargo. To determine the weight of the goods inside the container, subtract the tare weight from the gross weight. This information is usually stenciled on the container shell.

Tare weight is used by customs officials to determine taxes, tariffs, or tolls due on a shipment in transit. For emergency responders, knowing the gross weight of a container is useful when a damaged container needs to be lifted and moved. For safety reasons, lifting capacity should meet or exceed the gross weight of a loaded container. Knowing the tare weight of the goods is also useful in determining the personnel and resources needed to offload the freight container.

### ■ Freight Container Cargo Packaging Systems

The standard freight container can carry numerous types of hazardous and nonhazardous cargo. These can range from household goods and building materials to consumables, unrefrigerated nonperishable food, and almost every hazard class of Dangerous Goods. Freight containers have also been used for illegal purposes, including smuggling and human trafficking. (See Chapter 8.)

Emergency responders often refer to freight containers as a "surprise package." Beyond the standard markings and placards required by regulation on the intermodal freight container and shipping papers/manifests, what is inside a freight container may not be known until it is opened. Even then, it may take some time to identify the problem and the type of cargo involved.

Unless you work at a marine terminal or have one within your jurisdiction, the most likely emergency response scenario involving an intermodal freight container will be a highway transportation incident. Other possible scenarios might be a problem during container handling operations at a port or terminal facility, or as a result of a train derailment. (See **FIGURE 4-6** .)

Many intermodal freight container incidents involving hazardous materials or other dangerous cargoes are often due to inadequate securing of the cargo inside the container at the point of origin. Cargo that has not been properly loaded and secured can result in load shifts, making the freight container unstable in turns or during handling. Load shifts can also damage packaging or containers inside the freight container, thereby resulting in a breach of the container shell.

The most common categories of nonbulk packaging/ containers involved in freight container accidents involving hazardous materials are: 1) bags; 2) drums; 3) boxes; and 4) intermediate bulk containers (IBCs). The most common types of bulk packaging that may be involved in a freight container accident are both rigid and flexible intermediate bulk containers.

The following section provides an overview of these containers and packaging systems, and the most likely products and hazard class of the materials that may be shipped. Dealing with specific types of packaging and container emergencies will be discussed in Chapter 8.

**FIGURE 4-6** Intermodal freight container emergencies can involve a wide range of regulated and unregulated cargo and require many different skillsets to resolve.
Courtesy of Hildebrand and Noll Associates, Inc.

## Nonbulk Packaging

Nonbulk packaging will hold solid, liquid, or gaseous materials. The U.S. Department of Transportation (DOT) provides the following definitions:

- Liquid—Capacity of 119 gallons (450 L) or less
- Solid—Net mass of 882 pounds (400 kg) or less for solids, or capacity of 119 gallons (450 L) or less
- Compressed Gas—Water capacity of 1,001 pounds (454 kg) or less

Nonbulk packaging may consist of single packaging (e.g., drum, carboy, cylinder) or combination packaging—one or more inner packages inside of an outer packaging (e.g., glass bottles inside a fiberboard box, infectious disease sample containers). Nonbulk packaging may be palletized or placed in overpack containers for transport in vehicles, vessels, and freight containers. Examples include bags, boxes, carboys, cylinders, and drums.

### ■ Bags

Bags and sacks are designed to transport powders and granular materials. They typically hold 50 to 100 pounds (23 to 45 kg) of material. In the early days of cargo handling, bags and sacks were handled by manual labor. Today, robotics and mechanized handling systems have allowed nonbulk packaging to move cargo faster and more safely. This has also resulted in larger packaging which allows bag capacities to exceed 100 pounds. (See **FIGURE 4-7 A-C** .)

There are three general types of bags based upon their construction: 1) stitched bags; 2) folded and glued bags; and 3) shrink-wrapped bags. Bags have the following types of design and construction features:

- Flexible packaging constructed of cloth, burlap, kraft paper, plastic, or some combination of these materials. These packaging systems are normally fine for transporting and handling the cargo, but it makes them vulnerable to splits, tears, and punctures during accidents. These materials offer almost no fire protection for the cargo.
- Closed by gluing, heat-sealing, stitching, crimping with metal, or by twisting or tying.
- Typically contain up to 100 pounds of material, although large tote-style bags can be found that hold up to 500 pounds (226 kg). They may be palletized or hung inside of intermodal freight containers.

Bags can carry a wide range of materials and hazard classes, including explosives, flammable solids, oxidizers, organic peroxides, pesticides, and corrosives. (See **FIGURE 4-8** .)

**FIGURE 4-7** Bags can be stitched closed, folded and glued, or shrink-wrapped. They can be punctured, torn, or ripped open.
Courtesy of Hildebrand and Noll Associates, Inc.

By international ISO design standards, intermodal freight containers must be watertight to withstand multimodal transportation. Bags that are not sealed in plastic are often shrink-wrapped in plastic on a pallet. However, in a rollover or derailment scenario, water sensitive products can react if the freight container and internal pallet system is breached. Water reactive hazardous materials or moisture-sensitive organic materials can present significant problems for emergency responders when the bulk of the cargo remains inside the box.

Bags can also contain numerous other products that are not hazardous to people, animals, or the environment but can still present safety issues (e.g., slip and fall hazards) or minor health problems (e.g., nuisance dust) when the freight container is ripped open and the cargo is breached and released.

## ■ Drums

Drums are a cylindrical package constructed of metal, plastic, fiberboard, or other suitable materials. A top and bottom lid is welded or molded into place to seal the container.

The capacity of a standard drum is 55 gallons, although smaller or larger drums can be found ranging from 5 gallons to 55 gallons. A standard 55-gallon (208 L) liquid drum usually is filled to around 45 gallons to allow for product expansion.

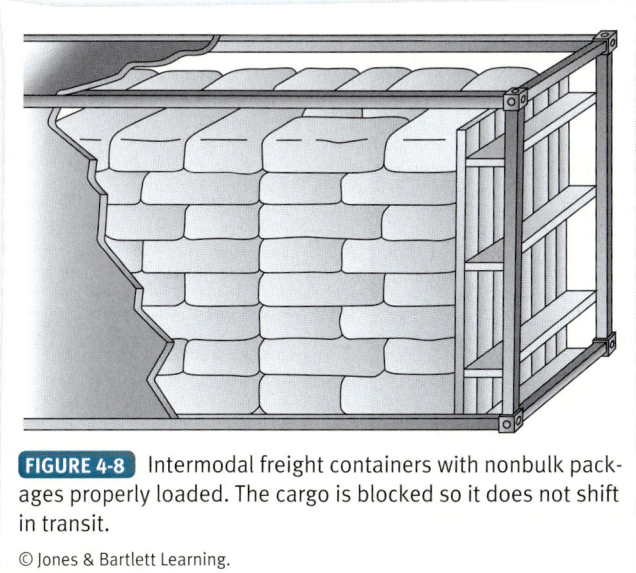

Drums can be open-head (i.e., the lid is removable) or closed-head (the lid is fixed and cannot be removed). Closed-head drums tend to be used more for hazardous materials, and usually contain two openings consisting of 2-in. and ¾-in. diameter plugs called bungs. Unsecure or untightened bungs are often the source of leaks, especially if the cargo has shifted in transit or the freight container is involved in a rollover. Chimes and seams can be the source of leaks on rusty metal drums.

Drums can contain liquids, solids, and various mixtures, including flammable and combustible liquids, flammable solids, oxidizers, organic peroxides, poisons, corrosives, radioactive materials, and hazardous waste.

As a general rule, although there are many exceptions, liquids are found in closed-head drums, and solids are found in open-head drums.

There are four general categories of drums that can be encountered in intermodal transit:

- **Steel Drums**—Commonly used for flammable and combustible liquids, poisons, mild corrosives, and liquids used in food production. Stainless steel drums are used for reactive or corrosive liquids such as nitric acid or oleum (super concentrate sulfuric acid @120-160% concentration). (See **FIGURE 4-9 A** .)
- **Aluminum Drums**—Used to hold materials that would react with rust or steel, and cannot be shipped in a poly drum. Contents are often combustible or toxic, such as pesticides. Caustic corrosives would *not* be shipped in an aluminum drum.

- **Plastic Drums**—Also known as poly drums. Commonly used for corrosive liquids, some flammable and combustible liquids, and food production liquids. (See **FIGURE 4-9 B** .)
- **Fiberboard Drums**—Commonly hold solid materials such as powder, granules, or pellets. May be toxic or corrosive, or may present little or no risk. They are usually lined, and may also hold gels and some low-hazard liquids. (See **FIGURE 4-9 C** .)

## Boxes

Boxes are commonly used as rigid outside packaging for other nonbulk packages. The inner packaging may be surrounded with absorbent materials.

Boxes are constructed of fiberboard, wood, metal, plywood, plastic, or other materials like fiberglass. Fiberboard boxes may contain up to 65 pounds (29 kg) of material; wooden boxes can contain up to 550 pounds (249 kg).

Almost any kind of material can be found inside a box, although they usually contain liquid or solid materials. Combination packaging using an outer box with inner packaging is commonly used for infectious disease samples and radioactive materials. (See **FIGURE 4-11** .)

Other types of containers that may be shipped inside boxes included carboys (glass or plastic) ranging in capacity to 20 gallons or aerosol cans.

## Bulk Packaging That May Be Placed on/in Transport Vehicles

The most common types of bulk packaging that may be placed inside an intermodal freight container are Intermediate Bulk Containers (IBCs). IBCs are divided into two categories: 1) Flexible Bulk Containers, and 2) Rigid Bulk Containers.

### Flexible Intermediate Bulk Container (FIBC)

FIBCs are also known as Super Sacks. These are typically a flexible sack or bag manufactured from woven fabric, cloth, or a plastic material with a capacity of as much as 2,000 pounds (907 kg) of product. They are commonly used for transporting dry products that will flow out of the sack, such as sand, fertilizer, powders, and granules of plastic. The sacks typically have a tied opening on the top and a bottom discharge. Loops or lift rings may also be installed on the top to facilitate movement by cranes or special handling equipment. Many are also permanently mounted on a skid for handling by forklift, while others may be transported on a removable skid.

A larger capacity variation of Super Sacks are Bulk Bags. These are typically made from polypropylene and can have a base dimension as large as 41 × 41 × 71 in.

**FIGURE 4-9** Metal drums may be constructed of steel or aluminum, (A) or may be made from plastic, (B) or fiberboard, (C)
Courtesy of Hildebrand and Noll Associates, Inc.

high with a capacity of 85 sq ft. The sacks are equipped with lifting loops and can be lifted with a single-point crane.

Another variation of the flexible container is heavy cardboard boxes containing an inner fabric or plastic bag. Once filled, the entire container may be shrink-wrapped to ensure container integrity, and secured to a pallet. These containers can be found transporting Class 9 (miscellaneous) hazardous materials. They may be found inside of intermodal freight containers, van trailers, and railroad box cars. They may also be found on some specially equipped flat cars and flatbed trucks that transport metal containers in permanently mounted racks. Depending upon the type of transport, these

containers may be restrained from movement by blocking and bracing. Highway dry vans commonly do not have any restraint added to prevent movement.

### ■ Rigid Intermediate Bulk Containers (IBCs)

Rigid Intermediate Bulk Containers (IBCs) are designed as a reusable industrial container for the transportation and storage of bulk liquid and granular substances. IBCs have a volume range somewhere between the standard 55-gallon drum and the typical transport tank (range of 145 gallons or 550 L to 792 gallons or 3,000 L), which is why they are called "Intermediate" Bulk Containers. IBCs typically carry chemicals, food ingredients, solvents, pharmaceuticals, or other materials.

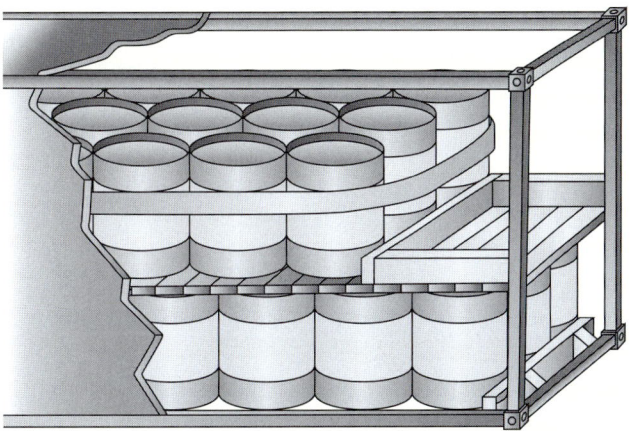

© Jones & Bartlett Learning.

Courtesy of Hildebrand and Noll Associates, Inc.

Courtesy of Hildebrand and Noll Associates, Inc.

**FIGURE 4-10** Drums present special cargo loading problems due to liquid surge inside the container. For stability they may be blocked and shrink-wrapped.

**FIGURE 4-11** Types of boxes and packaging.

Courtesy of Hildebrand and Noll Associates, Inc.

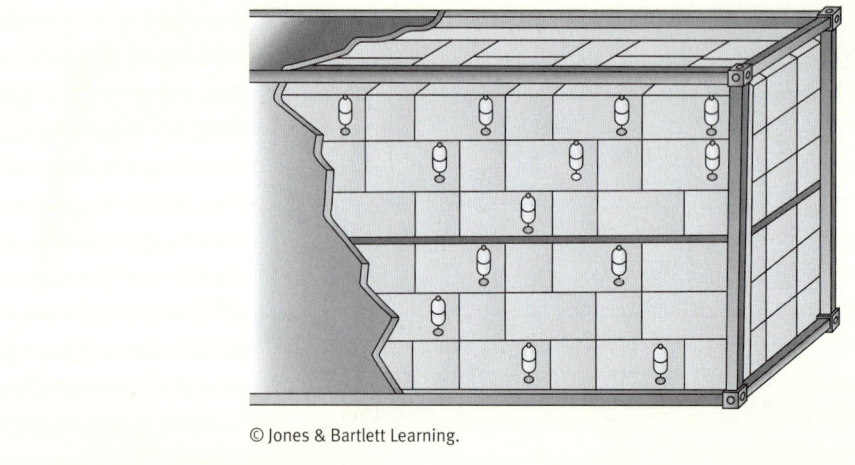

© Jones & Bartlett Learning.

Courtesy of Hildebrand and Noll Associates, Inc.

Courtesy of Hildebrand and Noll Associates, Inc.

**FIGURE 4-12** Boxes may be (A) blocked and (B) shrink-wrapped.

The key advantages of rigid IBCs over other types of containers are:

- They are stackable and mounted on a pallet that is integrated into the container system. They can easily be moved by a forklift or a pallet jack into or out of an intermodal freight container or in a warehouse. This allows the shipper or user to move capacities larger than the traditional 55-gallon drum.

- Discharge equipment can be built into the container, thereby reducing the need for handling.
- Depending upon the hazardous material and the container, IBC containers can be reused without cleaning. However, there are restrictions on the reuse of containers constructed of fabric materials.
- They reduce the costs associated with 55-gallon drum cleaning or disposal.

**FIGURE 4-13** Flexible Intermediate Bulk Container (FIBC).

Courtesy of Hildebrand and Noll Associates, Inc.

There are two basic types of IBCs:

- **All Metal Tank**—Circular or rectangular metal tank constructed with a top fill opening, and often with a bottom discharge piping or opening. Any safety devices (e.g., fusible plug) are found in the container fill lid. Also referred to as a tote, these containers are designed so that they may be stacked or moved with forklifts. (See **FIGURE 4-15** .)

    With a capacity of 300 to 400 gallons (1,135 to 1,514 L), they are commonly used for flammable liquids and solids. If involved in an emergency, leaks tend to be at the top fill opening. This opening often uses a locking ring secured by a bolt and nut, which may vibrate loose during transportation.

- **Polyethylene Tank and Steel Frame**—These containers consist of a polyethylene tank inside

**FIGURE 4-14** Flexible Intermediate Bulk Container (FIBC) Super Sack, (A) and fiberboard tote with flexible sack inside on a pallet, (B)

Courtesy of Hildebrand and Noll Associates, Inc.

**FIGURE 4-15** Types of all metal IBCs.

Courtesy of Hildebrand and Noll Associates, Inc.

**FIGURE 4-16** Polyethylene Tank and Steel Frame IBC.
© Greenshoots Communications/Alamy.

of a rigid steel frame. The tank is provided with a top fill opening and commonly has top discharge piping. In some instances, bottom discharge piping may be provided. Also referred to as a tote, these containers are designed so that they may be stacked or moved with forklifts. (See **FIGURE 4-16** .)

With a capacity of 300 to 400 gallons (1,135 to 1,514 L), they are commonly used for corrosive materials. In some emergencies, the polyethylene tank has separated from the steel framing, creating a mechanical stress and container failure.

## Summary

The most common intermodal container in circulation throughout the world is the intermodal freight container. There are an estimated 17 million intermodal freight containers in the world today, with a wide range of type, application, and use.

Freight containers easily move between the marine, rail, and highway transportation systems. In all three modes of transportation they can be found locked in place on the transport vehicle, stacked for marine and rail transportation, or stacked at ports and fixed facilities for short- or long-term storage.

The standard freight container is referred to as a "TEU," meaning that it is a Twenty-foot Equivalent Unit.

There are two general types of freight container construction: 1) corrugated-side containers, and 2) smooth-side containers. Corrugated-side freight containers are constructed of steel with sides consisting of a number of welded corrugated panels. Smooth-side freight containers are generally made of aluminum or fiberglass reinforced plywood (FRP). Supporting frames for all intermodal containers, freight containers, and portable tanks are built with corner fittings commonly referred to as corner castings. They are used to secure the container in rail and highway transit.

The gross weight of a container is the total weight of the container itself plus the maximum weight of the cargo that can be carried inside. The tare weight is the weight of the cargo inside the container.

The standard freight container can carry numerous types of regulated and unregulated cargo. Many intermodal freight container incidents involving hazardous materials or other dangerous cargoes are often due to inadequate securing of the cargo inside the container at the point of origin.

The most common categories of nonbulk packaging/containers involved in freight container accidents involving hazardous materials are bags, drums, boxes, and intermediate bulk containers (IBCs).

## References

1. EXSIF's Intermodal 407 Trailer Features. (June 25, 2015). Houston, Texas.
2. Noll, Gregory, G. and Michael S. Hildebrand, Glen Rudner, and Rob Schnepp. 2014. *Hazardous Materials: Managing the Incident*, 4th ed. Burlington, MA: Jones and Bartlett Learning. pp. 154–157.

# Nonpressure Intermodal Tank Containers

M.G.W. 30 480 KG
67 196 LB
TARE 7 550 KG
16 644 LB
CAP 16785 L T

# CHAPTER

# 5

## Chapter Outline

- Key Terms
- Introduction to Intermodal Tank Containers in General
- Design and Construction Features
- Tank Container Markings
- Types of Nonpressure Intermodal Tank Containers
- Inspections and Testing
- Summary
- References

## Key Terms

**Nonpressure Intermodal Tank Containers** Standardized 20-foot (6.058 m) stainless steel vessels for transporting liquids and solids, supported and protected by a steel frame that can easily be moved from ship to highway or rail transport vehicles and can be stacked for storage or transit.

## Introduction to Intermodal Tank Containers in General

The general classes of tank containers include nonpressurized (liquids and solids), pressurized (nonrefrigerated liquefied compressed gases), cryogenic liquid (refrigerated liquefied gases), and tube modules (nonrefrigerated compressed gases). **TABLE 5-1** summarizes the various types of containers.

This chapter will focus on nonpressure intermodal tank containers for liquids and solids (Hazard Classes 3–9) with T Codes 1 to 22 (IM 101 and 102 and IMO Type 1 and IMO Type 2) in the DOT Regulations. Intermodal tank containers for gases will be addressed in Chapter 6.

Nonpressure intermodal portable tank containers are standardized 20-foot (6.058 m) stainless steel vessels supported and protected by a steel frame. Note that there are some 40-foot containers in use. Like their counterpart, the intermodal freight container, they can easily be moved from ship to highway or rail transport vehicles and can be stacked for storage or transit. In the international shipping world they may also be referred to as tank containers, isotainers, or isotanks. (See **FIGURE 5-1**.)

The use of tank containers has increased greatly over the last 15 years. Factors influencing their acceptance include improved safety, portability, lower transportation costs, and the advantages of a multimodal transport system. According to a June 2013 survey of owners, leasing companies, and producers by the UK-based International Tank Container Organisation, there were an estimated 458,500 intermodal tank containers in circulation. Approximately 43,780 new tanks were manufactured in 2015.

Tank containers are well constructed and can remain in service for more than 20 years. Some tanks are sold for after-market use for transport or storage, although some are sold for scrap because of the value of stainless steel.

## Design and Construction Features

Like highway cargo tank trucks and railroad tank cars, there are different types of intermodal tanks built to many domestic and international standards. They are used to transport a varied and diverse range of commodities, including an increasing number of hazardous materials.

A nonpressure intermodal tank container usually consists of a single, noncompartmentalized vessel held within a sturdy, metal supporting frame that allows the unit to be lifted by appropriately designed handling equipment. Tank containers are built to the same

| Table 5-1 | Intermodal Tank Container Types with Corresponding Specifications by Regulatory Agency | | |
|---|---|---|---|
| **Intermodal Tank Container Types and General Uses** | **Current United Nations (UN) Specifications** *(adopted by DOT)* | **DOT Specifications** | **IMO Specifications** |
| **Nonpressure** *(used for liquid and solid hazardous materials)* | UN portable tank T Codes T1–T22 [1] | IM 101 portable tank [2] | IMO Type 1 |
| | | IM 102 portable tank [2] | IMO Type 2 |
| **Pressure** *(used for nonrefrigerated liquefied compressed gases)* | UN portable tank T Code T50 | Specification 51 portable tank [2] [3] | IMO Type 5 |
| | | Specification 60 steel portable tank [2] [4] [5] | |
| **Cryogenic liquid** *(used for refrigerated liquefied gases)* | UN Portable Tank T Code T75 | | IMO Type 7 |
| **Tube modules** *(used for nonrefrigerated compressed gases at high pressures)* | Multiple element gas containers (MEGCs). Specification is based on the type cylinder used in the tube module—3AX, 3AAX, or 3T. | | |

[1] T Code relates to specific products that can be transported as specified in regulations.
[2] DOT Specification 51, IM 101, or IM 102 tanks and IMO Type 1, Type 2, Type 3, and Type 4 tanks may not be manufactured after January 1, 2003; however, such tanks may continue to be used for the transportation of a hazardous material provided they meet the requirements of the current DOT regulations.
[3] A DOT Specification 51 portable tank can be used whenever a Specification 56, 57, or 60 portable tank is authorized.
[4] A DOT Specification 60 steel portable tank can be used whenever a Specification 56 or 57 is authorized.

Courtesy of Charles W. Wright.

standards as freight containers and can be handled the same way as all intermodal freight containers. The framing allows securement of the portable tanks on both vessels and surface vehicles.

## ■ Tanks Dimensions and Capacities

Nonpressure intermodal tank containers can be found around the world in various dimensions. There are no hard and certain "truths" about what you may encounter in terms of volume and length. The most common sizes are the 20 or 40 ft. in length or 8 or 8 ½ ft. high and 8 ft. wide; volumes do vary due to the commodity characteristics.

**FIGURE 5-1** Intermodal tank containers may be found in marine, highway, rail, and in rare cases air transportation modes.
Courtesy of Hildebrand and Noll Associates, Inc.

Similar to railroad tank cars, there is a common process for identifying the sides and ends of an intermodal tank container. Tank containers are usually described in relation to the tank end that is fitted with the discharge valve; this is referred to as the discharge end of the tank. Facing the discharge valve at the end of the tank, the right side of the tank would be to the right. If the container is not equipped with a discharge valve, the container or the frame is usually marked "Front" and "Rear." (See **FIGURE 5-2** .)

The following represents the most common dimensions and tank capacities.

- Length—The standard tank container is 20 ft. (6.058 m) long, however, they may also be found in 10 ft. (2.991 m) and 40 ft. (12.192 m) lengths. In the international tank industry, 95% of all tanks built are 20 ft. long. *Note:* 30-foot containers were originally considered an ISO standard size; however, the size never caught on since it was not easily interchangeable with the common 20-foot and larger 40-foot sizes. Very few American companies used them.
- Width—The tank is 8 ft. (2.438 m) wide.
- Height—Generally 8 ft. (2.438 m) and 8 ft. 6 in. (2.591 m) and 4 ft. 3 in. (1.296 m) tall.
- Volume—Generally ranges from 2,377 gallons (9,000 L) to 7,000 gallons (26,497 L). Volume of the 40-foot tank is 7,123 gallons. Most nonpressure tank containers are 6,340 gallons or less.

## ■ Construction Features

Tank containers can be described by three primary constructions features: 1) tank shape, 2) materials of construction, and 3) type of tank frame.

**Container Orientation**

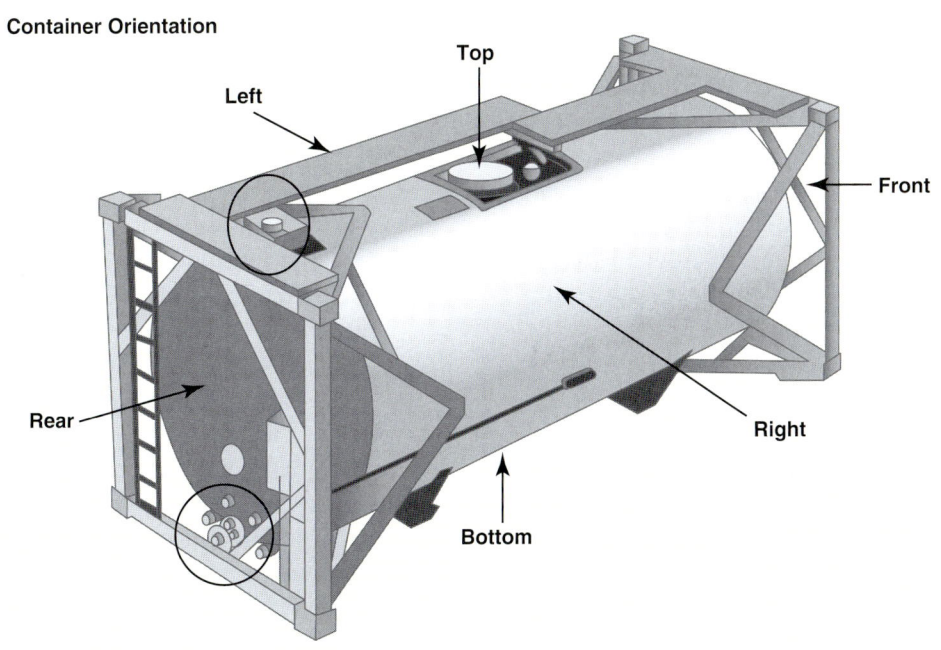

- **Tank Shape**—The tank itself is generally built as a cylinder enclosed at the ends by heads. Although rare, rectangular tanks may also be found. The intermodal tank is usually a single, noncompartmentalized vessel whose capacity does not exceed 6,340 gallons (24,000 L). Multicompartment intermodal tanks are rare; however, when found, each compartment is constructed as a separate tank. (See **FIGURE 5-3**.) Although there are typically no internal baffles (surge plates) on 20-foot intermodal portable tanks, they may be found on 40-foot portable tank containers.

Tank containers may be equipped with various features, including the following:

- **Linings**—To protect the tank from its contents. Linings can range from rubber, glass, and other coverings applied to the inside of the tank after it is built.
- **Refrigeration units**—Most tanks with refrigeration units are not provided with their own power source, and an external supply source is required. Depending upon the mode of transportation, options may include a ship's power system, a plug-in at a fixed facility, a chassis-mounted generator for highway use, a freight car generator, or a generator container.
- **Heating units**—For handling very viscous products. Options include electrical or steam

heating. Electrical heater coil units are commonly found where the product must be heated during transportation, and may be installed either internally or externally. They operate on either 200-240 volt or 340-480 volt, 3 phase electrical supply. Steam heating is provided by pumping steam through external heater coils on the lower

**FIGURE 5-3** Double-compartmented tank container is indicated by the marking "Compartment 2". Note the location of the foot valve.

Courtesy of Hildebrand and Noll Associates, Inc.

half of the tank. According to Brassington's *Safe Handling of Containers" Safety Briefing Pamphlet Series #30*, "The principle means of applying the heat is by steam, hot water or hot glycol being piped through heating coils welded directly onto the exterior of the barrel."

- **Insulation**—To moderate the effects of ambient temperature on the contents. Insulating materials include polyurethane foam, polystyrene foam, mineral wool, and fiberglass. Insulation is usually 3 to 4 in. thick, and is always covered with a jacket (also known as cladding) with flashing to make it weather resistant. Jackets are made of metal, at least 1 mm thick, or an equivalent thickness of plastic reinforced with either glass or fiber. Remember that the tank is attached to the container framing; the insulation is not an integral part of the tank.
- **Electrical controls**—The electrical control box is mounted on the tank frame at the rear of the tank container (i.e., the end where discharge valve is located). The control box will contain fuses or circuit breakers, temperature controls, a main switch to isolate the tank container from the power supply, and a method of selecting the correct circuit for the available main power supply.

- **Materials of Construction**—Because of its strength and excellent properties in cold temperatures, over 90% of intermodal tanks are constructed of stainless steel. The remainder are constructed of mild steel. Aluminum and magnesium alloy tanks may also be found, but they cannot be used in marine transportation.

Shell thickness is measured in terms of "equivalent thickness in mild steel" after forming. For shell thickness requirementthe minimum thickness in terms of mild steel is 6 mm, although some nonpressure intermodal tank containers have 8 and 10 mm minimums (T20-T22). (See the table in the federal regulations 172.102(7) - TABLE OF PORTABLE TANK T CODES T1-T22 for more information.)

If you see external rings on an intermodal tank, it is most likely a single-shell, stainless steel tank. Most tanks are constructed to the pressure-vessel standards of the American Society of Mechanical Engineers (ASME). Welds are x-rayed, while the welds on carbon steel tanks are post-weld stress relieved.

- **Type of Tank Frame**—The supporting frame of a tank container protects the tank and allows for stacking, lifting, and securing the container. It also supports the walkways and ladders. The most common size supporting frame for tank containers is 20 ft. long, 8 ft. wide, and 86 in. high. Very few tank containers used within the United States are longer than 20 ft., but internationally they can be as long as 40 ft.

There are three main types of framework configurations found on nonpressure intermodal tank containers: 1) beam, 2) frame, and 3) collar. All three designs can be top lifted, are stackable, and the tank barrel as well as all of the fittings and attachments must remain within the envelope of the tank frame (i.e., none of the parts can protrude past the faces of the corner fittings). This strict design and construction feature makes tank containers less susceptible to damage.

- **Box Frame**—Encloses the tank in a cage-like framework with continuous side rails. Two end frames are separated by two main beams which form the box type. Often the lower beans are castellated, a manufacturing method of lightening the main beams by cutting holes to reduce the tare weight, which results in an increase of payload. The rails are made of lightweight metal construction and do not add much in structural strength; they are there mainly to support the walkway. (See FIGURE 5-4.)
- **Beam Frame**—Uses four load-bearing structural supports which are attached to the end frames and attached to the tank barrel at four

**FIGURE 5-4** The box type frame of a tank container protects the tank and allows for stacking, lifting, and securing the container.

Courtesy of Hildebrand and Noll Associates, Inc.

| Material of Construction | Regulated Commodities | Nonregulated Commodities |
|---|---|---|
| Stainless Steel Tanks | 0.1875 in. | 0.125 in. |
| Steel Tanks | 0.375 in. | 0.25 in. |

**FIGURE 5-5** A. The beam frame uses four load-bearing structural supports attached to the end frames and to the tank barrel at four locations around the circumference of the tank. Note the "40 T6" designation and discharge valve at the rear of the tank.

Courtesy of Hildebrand and Noll Associates, Inc.

locations around the circumference of the tank. The beam frame supports are attached to plates that are welded to the tank to ensure load sharing and act as a barrier between carbon steel and the stainless steel components. The "beam type" design relies upon the inherent strength of the tank as a beam. (See **FIGURE 5-5**.)

• **Collar Frame**—The simplest type of tank designs. The key distinguishing feature is the minimum number of contact points between the frame and the tank barrel. Another key feature is the minimum number of differing metal materials in its construction. The tank barrel is

attached to the end frames by using a stainless steel collar that is welded to the end dome. The collar is continuous at the front (nondischarge end). At the discharge end of the tank, some designs have a break in the collar where the discharge valve is located. (See **FIGURE 5-6**.)

• **Corner Castings**—Like freight containers, supporting frames for tank containers are built with corner fittings commonly referred to as corner castings. They are used to secure the tank and lift it with standard container handling equipment. Cast iron corner castings are prohibited. In the event of an accident, the corner castings may be used for lifting or moving the tank, but only after consultation with the tank's owner or manufacturer.

## Tank Container Markings

There are a number of markings on tank containers that can be used to gain knowledge about the tank. These markings include the following:

■ **Owners Identification Code and Number**—Tank containers are registered with the International Container Bureau in France. They must be marked with the owner's code and serial number when talking about tank containers. The initials indicate ownership of the tank, and the tank number identifies the specific tank. These markings are generally found on the right-hand side of the tank (facing it from either side) and on both ends. (See **FIGURE 5-7**.) They may be displayed either on the tank itself or on the tank frame.

■ **Specification Marking**—The specification marking indicates the standards to which a portable tank

**FIGURE 5-6** The collar frame is the simplest form of design. The container is attached to the end of the frames. It can easily be identified by the lack of a horizontal structural support on the frame.

Courtesy of Hildebrand and Noll Associates, Inc.

**FIGURE 5-7** This intermodal portable tank container has markings and placards that convey a lot of information. The owners identification number TRU 023319 1 can be seen in the first three letter "TRU" in the upper right hand corner. The letter "U" indicates the type of service. Also note the overhead electrical hazard warning. Below the owner's identification code is the UN 22T6 design specification. The total capacity in U.S. gallons is indicated as 6.340 gallons. The four digit UN Number 3109 indicates it is an organic peroxide liquid. The product is also corrosive and flammable.

Courtesy of Hildebrand and Noll Associates, Inc.

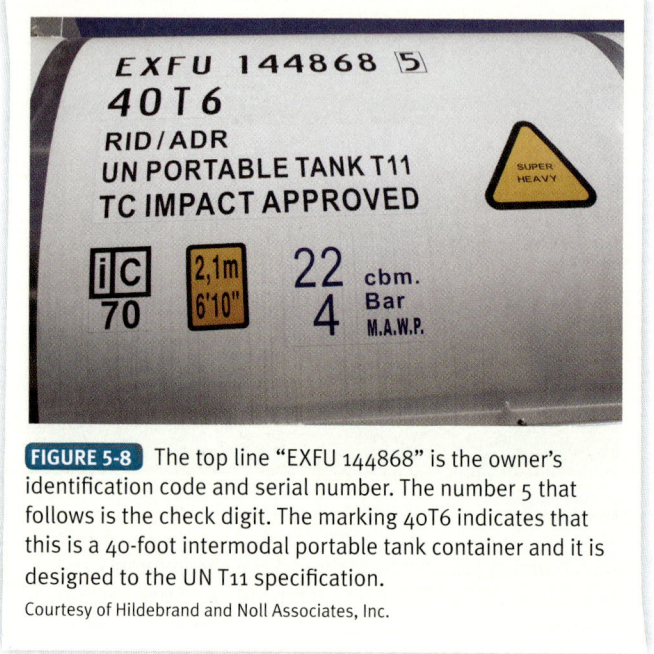

**FIGURE 5-8** The top line "EXFU 144868" is the owner's identification code and serial number. The number 5 that follows is the check digit. The marking 40T6 indicates that this is a 40-foot intermodal portable tank container and it is designed to the UN T11 specification.

Courtesy of Hildebrand and Noll Associates, Inc.

was built. Tank containers must meet international specifications. See Table 5-1 for the list of specifications that could be found on nonpressure intermodal tank containers—remember the specification is UN portable tank and may have the T code associated with it. These markings will be on both sides of the tank, generally near the tank owner's identification code and serial number. (See **FIGURE 5-8** .)

- **Special Permit Exemption Marking and Five-Digit Number**—Exemptions are sometimes authorized from DOT regulations. In these cases, the outside of each package/container must be plainly and durably marked "DOT SP" followed by the exemption number assigned (e.g., DOT SP8623). On intermodal tanks, these markings must be in 2-in. letters.
- **AAR-600 Marking**—For interchange purposes in rail transportation, intermodal tank containers should conform to the requirements of Section 600—"Specification for Acceptability of Tank Containers" of the Association of American Railroads (AAR) Specifications for Tank Cars. The primary requirements of AAR 600 are shown in **TABLE 5-2** below.
- **Minimum Tank Design Pressure**—The minimum tank design pressure is 35 psi, but in no case less than the vapor pressure of the commodity at 115°F (46.1°C). Tanks meeting these requirements will display the "AAR 600" marking in 2-in. letters

on both sides near the tank's reporting marks and number The "AAR 600" marking indicates tanks that can be used for regulated materials, while the "AAR-600NR" marking indicates tanks that cannot be used for regulated materials.

- **Country, Size, and Type Markings for Tank Containers**—The tank will display a size/type code. The country code (two or three letters) indicates the tank's country of registry. The four-digit size/type code follows the country code. The first two numbers jointly indicate the container length and height. The second pair of numbers is the type code, which indicates the pressure range of the tank.

Common Size Codes
20 = 20 ft. (8 ft. high)
22 = 20 ft. (8 ft. 6 in. high)
24 = 20 ft. (> 8 ft. 6 in. high)

Common Type Codes—
Maximum Allowable Working Pressure
Nonhazardous Commodities (liquid or solid low-hazard or nonhazardous materials)

| Table 5-2 | Minimum Shell Thickness | |
|---|---|---|
| Carbon Steel | nonregulated materials | 0.25 in. (6.35 mm) |
| Carbon Steel | regulated and nonregulated materials | 0.375 in. (9.52 mm) |
| Stainless Steel | regulated materials | 0.164 in. (4.18 mm) |
| Stainless Steel | nonregulated materials | 0.118 in. (3.0 mm) |

T0 = < 0.44 (6.4 psig) Bar test pressure
T1 = 0.44 (6.4 psig) to 1.47 (21.3 psig) Bar test pressure
T2 = 1.47 (21.3 psig) to 2.94 (42.6 psig) Bar test pressure
T3 = spare

Hazardous Commodities (for liquid and solid hazardous materials)
T4 = < 1.47 (21.3 psig) Bar test pressure
T5 = 1.47 (21.3 psig) to 2.58 (37.4 psig) Bar test pressure
T6 = 2.58 (37.4 psig) to 2.94 (42.6 psig) Bar test pressure

Hazardous Commodities (gases)
T7 = 2.94 (42.6 psig) to 3.93 (57.0 psig) Bar test pressure
T8 = > 3.93 (57.0 psig) Bar test pressure
T9 = spare

- **Data Plate**—Technical details, approval, and operational data can be found on the data plate. (See **FIGURE 5-9**.) Each tank must have a corrosion-resistant data plate permanently affixed to the portable tank in a location that is readily accessible for inspection.
The following information is required:
  - Country of manufacture
  - U N
  - Approval country
  - Approval number
  - Alternative Arrangements "AA"
  - Manufacturer's name or mark
  - Manufacturer's serial number

**FIGURE 5-9** The data plate contains operational data pertaining to the tank. This data plate indicates that the tank was constructed to UN portable tank container specification T11.

Courtesy of Hildebrand and Noll Associates, Inc.

- Approval agency (*Authorized body for the design approval*)
- Owner's registration number
- Year of manufacture
- Pressure vessel code to which the shell is designed
- Test pressure____bar gauge
- MAWP____bar gauge
- External design pressure (*not required for tanks used for refrigerated liquefied gases*)____bar gauge
- Design temperature range____°C to____°C. (*For tanks used for refrigerated liquefied gases, the minimum design temperature must be marked.*)
- Water capacity at 20 °C/____liters
- Water capacity of each compartment at 20°C ____liters
- Initial pressure test date and witness identification
- MAWP for heating/cooling system____bar gauge
- Shell material(s) and material standard reference(s)
- Equivalent thickness in reference steel____mm
- Lining material (when applicable)
- Date and type of most recent periodic test(s)
- Month____Year____ Test pressure____bar gauge
- Stamp of approval agency that performed or witnessed the most recent test
- For tanks used for refrigerated liquefied gases:
  - Either "thermally insulated" or "vacuum insulated"____.
  - Effectiveness of the insulation system (heat influx)____Watts (W).
  - Reference holding time____days or hours and initial pressure____bar/kPa gauge and degree of filling____in kg for each refrigerated liquefied gas permitted for transportation.

Current regulations require that the tank container data plate indicate the water capacity of each compartment at 68°F (20°C). The standard 20-foot tank container is a single-compartmented tank and is generally not equipped with a baffle, however, some low-hazard and nonhazardous materials will have baffles or surge plates. The 40-foot container may include an internal baffle. While rare, you may find 40-foot tank containers with more than one compartment. The data plate indicates if the tank is divided into compartments but does not require that the number of compartments in the tank be listed. As of 2013, there were no current regulations to mark 40-foot containers to indicate if they are fitted with a baffle unless the compartment is divided by baffles into sections of not more than 1,981 gallons (7,500 L). In this case, the tank identification plate must add the letter "S" in their respective capacity sections.

- **CSC Plate**—A Container Safe Convention (CSC) plate is required to be fastened to each tank container used in international intermodal service, typically as part of or adjacent to the data plate. The CSC Plate should be somewhere around the data plate on the rear of the tank. Each CSC plate contains information about the container including:
  - Words "CSC SAFETY APPROVAL"
  - Country of approval and approval reference
  - Date (month and year) container was manufactured
  - Manufacturer's identification number of the container
  - Maximum gross weight
  - Allowable stacking weight for 1.8 g
  - Racking test load value
  - Next examination date or Approved Continuous Examination Program reference number

  The plate may include the owner of the shipping container and their contact information, but if not, this information can be obtained using the manufacture identification number.

- **Customs Approval Plate**—Customs regulations requires all containers that are used in international intermodal service to carry a customs approval plate, decal, or three-letter owner identification code registered with BIC. The customs approval plate will display the following information:
  - Words "Approved for Transport Under Customs Seal"

- Customs approval number
- Type container
- Manufacturer's number (a means to identify the owner and their address)

- **Document Tube**—A document tube is intended to contain the tank's shipping documents, cleaning documents, or a Safety Data Sheet (SDS). Information found in these tubes may be out of date. Constructed of metal or plastic, the tubes are normally at or near the rear end in proximity to the discharge valve. It may be found in proximity to the data plate. (See **FIGURE 5-10**.)

- **Tank and Valve Test Dates**—If installed, tank and safety valves must have a retest interval no greater than five years. Retest and test due dates must be marked or stenciled on either the tank or on the data plate.

- **Hazardous Materials Markings and Placards**—Tank containers containing regulated materials must be marked and placarded. (See **FIGURE 5-11**.) The four-digit identification number must be displayed. Placard size (as required by 49 CFR 172.519) must measure at least 9.84 in. on each side that conforms to international placard sizes, which might eliminate the requirement for multiple placarding. Around busy international port facilities you may also see nonstandard size placards.

  Except for Class 7 and DANGEROUS placards, text indicating the hazard is not required. Hazard class or division number must be shown.

  For domestic shipments within the United States, DOT regulations require that the proper shipping name be printed in two-in. letters on two opposing sides of the container shell. In some

**FIGURE 5-10** (A) The document tube contains the tank's shipping papers, cleaning documents, or Safety Data Sheet. Unscrew the cap to gain access. (B) Safety Data Sheet.

Courtesy of Hildebrand and Noll Associates, Inc.

**FIGURE 5-11** This portable tank container is placarded Corrosive and has a Marine Pollutant warning.
Courtesy of Hildebrand and Noll Associates, Inc.

**FIGURE 5-12** Tank containers containing regulated materials must be marked and placarded when required. Note that this container shows the Corrosive placard with the ADR/RID markings.
Courtesy of Hildebrand and Noll Associates, Inc.

cases, you may see the synonym of the product marked on the container as well. On international shipments, both foreign and U.S. placards are sometimes found. European shipments may also carry the ADR/RID Markings. (See **FIGURE 5-12** .)

Some specific situations which responders should be aware of regarding the placarding of international shipments are as follows:

- For domestic shipments, tanks loaded with a regulated commodity must display the appropriate DOT placard to correspond with the classification of the commodity. In addition to the required DOT placard, any additional placard authorized by the International Maritime Dangerous Goods Code (IMDG) should be found. (See 49 CFR 172.502(c)(1)(.b).)
- The United Nations Globally Harmonized System of Classification and Labeling of Chemicals (or GHS) is an internationally recognized system. It is designed to replace the various classification and labeling standards used in different countries by using consistent criteria for labeling and placarding. As GHS is phased in, responders will encounter differences in appearances of labels and placards on international shipments.

## Types of Nonpressure Intermodal Tank Containers

Tank containers are classified according to the specification of the portable tank and its fittings. The tank container class determines which products may be transported. The general classes of tank containers include nonpressurized, pressurized, and specialized.

The widely used specification tank containers permitted to transport nonpressure intermodal tank containers for liquid and solid hazardous materials in North America include (see information from Table 5.1) UN portable tanks T1 to T22, DOT IM 101 and DOTIM 102, and IMO Type 1 and IMO Type 2.

## ■ Nonpressure Intermodal Tank Containers

Nonpressure bulk liquid tank containers may have a stainless steel or mild steel tank shell, be insulated or uninsulated, have heating or cooling equipment, or be lined with a protective coating.

Although classified as "nonpressurized", these containers can have a working pressure up to 100 psig—which can present safety hazards if valves, and attachments are not handled properly. Nonpressure tank containers comprise over 90% of the total number of tank containers, with the most common being the IM 101 and IM 102 portable tank containers. The only method to distinguish an IM 101 from an IM 102 container is to physically inspect the data plate or container markings. Nonpressure tank containers can transport both liquid and solid materials at maximum allowable working pressures (MAWP) of up to 100 psig (6.8 bar). Tanks are tested to at least 1.5 times the MAWP. Higher pressures for some liquids (e.g.,T15–T23) can be found. The T20 and T21 tanks have a 10 bar MAWP.

- **IM 101 Portable Tanks (International—IMO Type 1)**—These tanks are built to withstand MAWPs ranging from 25.4 (1.75 bar) to 100 psig (6.8 bar). An ASME certification or stamp is not a requirement.

IM 101 tanks are used for transporting both hazardous and nonhazardous liquid and solid materials, including toxic, corrosive, and flammables with flash points below 32°F (0°C). Capacities are normally in the range of 5,000 to 6,300 gallons. (See **FIGURE 5-13** .)

- **IM 102 Portable Tanks (International—IMO Type 2)**—These tanks are built to withstand lower MAWPs ranging from 14.5 psig (1.0 bar) up to but not including 24.4 psi (1.75 bar). An ASME certification or stamp is not required.

  IM 102 tanks transport liquid and solid materials such as whiskey, alcohols, some corrosives, pesticides, insecticides, resins, industrial solvents, and flammables with flash points ranging from 32°F (0°C) to 140°F (60°C). These containers are also commonly used for the transport of nonregulated materials, such as food-grade commodities. Capacities are normally in the range of 5,000 to 6,300 gallons. (See **FIGURE 5-14** .)

- **Nonpressure Tank Container Fittings**—The following fittings can be found on the nonpressure tank containers to make them both safe and functional. Although many of these fittings are similar to those found on a highway cargo tank truck, container fittings and threads are normally British Standard Pipe (BSP). British Standard Whitworth (BSW) are used on the manhole wing nuts. See **FIGURE 5-15** .

- **Access to Container Tops**—Tank containers will have a ladder built into the rear frame. Some of these are actual ladders while others are more of a climbing frame. Both present fall hazards.

  Access to top fittings is normally via a vertical climbing ladder and topside walkway installed on the

**FIGURE 5-14** IM 102 tanks are used for the transport of non-regulated materials, such as food-grade commodities. The only way to determine if the tank is an IM 102 is to look at the data plate attached to the tank.

container. If access to the top of the container is required for loading and offloading, a decal warning of potential overhead electrical hazards will be marked as shown in **FIGURE 5-16** . Overhead electrical hazards may or may not be present, but this serves as a warning to do an overhead safety check before climbing on top of or moving the container.

Tank container ladders may be loose or weakened or missing rungs/steps. During emergency situations, emergency responders should use portable ladders that can be properly butted and secured for safe access to the top of the container. When dealing with emergencies that require responders to go topside, always have a second way off of the tank. Do not walk on the tank shell. Remember that a fall from a height of 8 ft. can cause serious injury or death. (See **FIGURE 5-17** .)

Tank container fittings on the top of the tank container include the following:

- **Spillbox**—On most nonpressure tanks, the top fittings are surrounded by a spillbox that protects the shell of the tank from product spillage. The spillbox chamber is sometimes fitted with a manhole and a hinged cover for accessing the inside of the tank for tank cleaning. Spilled materials, as well as rainwater in the spillbox, are drained away to the ground through one or more small open tubes. Many false incidents have been reported when product was thought to be leaking from the tank when it was simply rainwater or condensate dripping from the drain tube. That being said, every leak should be considered to be an actual product leak until it is proven otherwise. (See **FIGURE 5-18** .)

**FIGURE 5-13** IM 101 tanks are used for transporting both hazardous and nonhazardous materials, including toxic, corrosive, and flammables with flash points below 32°F (0°C).

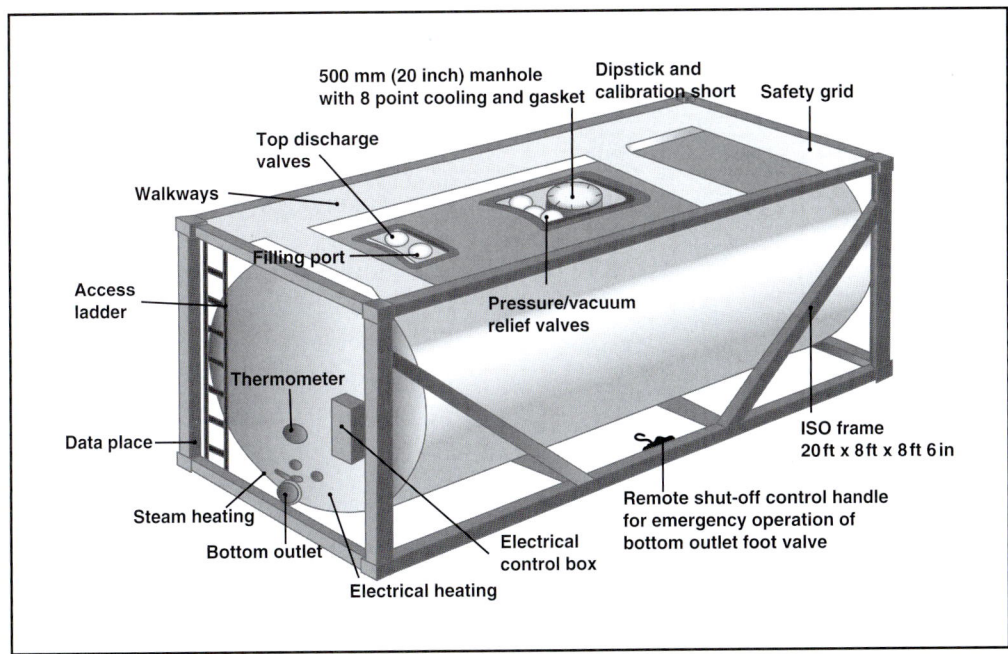

**FIGURE 5-15** Various types of fittings can be found on intermodal tanks.

© Jones & Bartlett Learning.

■ **Manhole Cover and Dipstick**—An 18- to 22-in. (45.7–55.8 cm) manhole is located on top of the tank at the center. It is enclosed by a hinged or bolted cover fitted with six or eight large wing nuts. A replacement gasket is used with the manhole cover, with neoprene (food quality) being standard. Other gasket materials may include Viton® and Teflon®. Some companies require the replacement of gasket materials once the manhole cover is opened, particularly when using "soft" gasket materials (e.g., Teflon®). For example, if a manhole cover is opened and the gasket is Teflon®, the gasket should be replaced and not reused. (See **FIGURE 5-19**.)

A dipstick may be either inside the manhole or laying within the spillbox. It is used in conjunction with a calibration chart, also known as a strapping chart, to measure the amount of product in the tank. Tanks must be loaded to no more than 80% of capacity to avoid sloshing. For liquids, a

**FIGURE 5-16** Various types of safety warning signs can be found on intermodal tanks. This marking warns operators to check for overhead electrical hazards before climbing on top of the tank.

Courtesy of Hildebrand and Noll Associates, Inc.

**FIGURE 5-17** This tank includes a "Do Not Walk on Tank Shell" marking at the end of walking surface. Do not walk on any tank shell; they are slippery. A fall from a height of 8 ft. (2.4 m) can cause serious injury or death. Note the corrugated drip tube.

Courtesy of Hildebrand and Noll Associates, Inc.

**FIGURE 5-18** Topside fittings are surrounded by a spillbox protecting the shell of the tank from product spillage. A drip tube drains a spill or rain water through to the bottom of the tank. A dipstick can be seen in the upper left. The pressure relief device is shown between the dipstick and the manhole cover to the left. The manhole cover can be a source of leaks if the container is involved in a rollover.

Courtesy of Hildebrand and Noll Associates, Inc.

minimum outage of 2% of the total capacity of the tank must be provided in rail transportation.

- **Top-Loading Valves**—Top-loading valves are attached to a removable eduction pipe (e.g., dip leg, dip tube, or siphon tube) running into the tank. They can range from 1-1/2 to 4 in. in diameter, although it is typically a 3-inch diameter ball or butterfly valve with a four-bolt flange. If a blind flange is in place for transportation, fittings for transfer operations must be provided by the shipper or the tank supplier. A sump is sometimes provided at the bottom of the eduction pipe for maximum discharge capability.

- **Foot Valves**—Foot valves are used to remove product from the tank by connecting the valve to a discharge hose. Foot valves can range from 1-1/2 to 4 in. in diameter. When a tank container is intended to transport hazardous materials, a three serial device configuration is required. This consists of two externally operated, bottom-outlet valves and a liquid tight closure. Typical designs are for a 3-in. internal foot valve, a 3-in. external butterfly

**FIGURE 5-19** (A) The manhole cover is in the open position. (B) This topside configuration shows the manhole cover unsecured. The dipstick is located at top-center. The pressure relief device is shown in the upper left.

Courtesy of Hildebrand and Noll Associates, Inc.

valve, and a 3-in. British Standard Pipe (BSP) thread screw cap. They are connected in series with a replaceable gasket between them. Some containers may also have electrical sensor connections adjoining or as part of the bottom outlet valve to provide for product identification and product overfill protection. This is similar to the sensor connection commonly found on gasoline tank trucks to prevent product overfills at loading racks. (See FIGURES 5-20.)

A liquid-tight closure on the external valve is also required. It may be a blind flange, a screw cap, or a cam-lock cap attached to the external valve. Blind flanges are required for international shipments. AAR 600-12 requires a positive lock on the external bottom valve to lock in the closed position.

Emergency response experience shows that most leaks occur at the blind flange as compared to the valve itself. Flange leaks can be easily controlled by tightening down on the flange bolts.

When three shutoff devices are required, the first bottom outlet (foot valve) must be equipped with a remote shut-off device. This emergency device can close the internal bottom outlet valve (foot valve) from a remote location. The device consists of a steel cable, often plastic coated stainless steel, or a rod that leads from the foot valve along the side of the container. The action of pulling the wire or rod will lift the foot valve operating handle over the cam position, thereby snapping the internal valve closed. (See Figure 8-6 in Chapter 8.)

Prior to the introduction of the second generation UN portable tank containers in the 2001–2002 timeframe, international dangerous

FIGURE 5-20 When a tank container is intended to transport hazardous materials, a three serial device configuration is required.

Courtesy of Hildebrand and Noll Associates, Inc.

goods regulations did not require a remote shutoff device. Responders may encounter first generation tanks in circulation that have not been retrofitted with a remote shutoff device.

As you face the discharge end of the tank, the emergency shut-off can be on either the right-hand side or left-hand side near the far end. It is usually a cable-actuated device, although hydraulic or pneumatic devices may be found. (See **FIGURE 5-21**.) Fusible links or nuts may be found on cable-actuated remote shut-off devices. In the event of a spill fire under or around the tank container, the fusible device will melt, releasing cable tension. Fusible links are required to actuate at temperatures not greater than 250°F (121°C).

- **Air Line Connection**—An air line connection can be used for pressure unloading, vapor return, and blanketing the contents with an inert gas. It is normally a 1-1/2-in. line. In some instances, a 1-1/2-in. ball valve and pressure gauge may be found. This connection can be found on the tank's top, normally within the spillbox. If a blind flange is in place for transportation, fittings for transfer operations must be provided by the shipper or the tank supplier. When not in use, air lines must be blanked or capped off to prevent accidental operation of the valve. (See **FIGURE 5-22**.)

- **Steam Line Connection**—Some tank containers carry cargo that needs to be heated so that it can be removed from the tank. These tanks are equipped with either an electrical or steam heating system,

**FIGURE 5-22** Air line connections are used for pressure unloading, vapor return, or blanketing the contents with an inert gas.
Courtesy of Hildebrand and Noll Associates, Inc.

most likely equipped with a thermometer. (See **FIGURE 5-23**.) The thermometer may be either an immersion sensor or a surface sensor connected to a temperature gauge. Temperature gauges will usually read both °F and °C.

- **Discretionary Fittings**—Beyond what is required under regulations, there are a variety of fittings that a tank owner may add. These can include liquid level indicators, low- and high-level alarms, and hydraulic or pneumatic operating systems. The IMDG Code prohibits sight glasses as a method of checking liquid levels.

**FIGURE 5-21** This emergency shutoff device is activated by grabbing the breakoff plug and snapping it off. This releases the internal pressure in the line and allows the internal valve to close automatically.
Courtesy of Hildebrand and Noll Associates, Inc.

**FIGURE 5-23** The steam line connection can be seen in the lower right hand corner.
Courtesy of Glen Rudner.

**FIGURE 5-24** Tank containers that carry Hazard Class-3 flammable liquids that can accumulate and discharge static electricity are equipped with bonding and grounding connections.

Courtesy of Tyler Bones.

- **Grounding Connections**—Tank containers that carry Hazard Class-3 flammable liquids that can accumulate and discharge static electricity are equipped with bonding and grounding (earthing) connections. These tank containers should be grounded and bonded during loading, offloading, or tank cleaning operations. (See **FIGURE 5-24** .)

## ■ Tank Container Safety Devices

The following safety devices can be found on the IM 101 and IM 102 tank containers. They include the following:

- **Pressure/Vacuum Relief Valves**—These are generally found in pairs on nonpressure tank containers. Typically, two 3-in. diameter, spring-loaded devices are installed on top of the portable tank near the manhole. (See Figure 5-18 above.) A combination pressure/vacuum relief device protects the tank from both over-pressure and a vacuum of more than 0.75 psig negative pressure. The valve will be marked to

indicate its settings. In many cases, the pressure relief valve will also have a rupture disc (i.e., burst disc) located between the safety relief valve spring and the commodity to protect the spring from the commodity. Containers transporting flammable liquids will have a flame-arresting screen on the pressure vacuum ports on the valve.

This fitting may also be found with a pressure gauge or "telltale valve" to determine if the disc is ruptured. (A telltale valve is a valve to atmosphere between two blocking valves that can be periodically opened to see if one or both of the other valves is leaking by as evidenced by pressure buildup or smell depending on which side is leaking.) The pressure gauge normally reads in both psi and bars (1 bar = 14.5 psi); the gauge should always read zero. Responders may find situations where the burst disc has failed but the relief valve has not actuated. Possible reasons may include container overfilling, product expansion due to ambient heating, hydraulic surge effects, etc.

## Inspections and Testing

### ■ Inspections

Under DOT regulations (49 CFR 173.32c(d)), a pre-trip inspection of the intermodal tank is required. This is primarily a visual external inspection to spot mechanical damage, deterioration, equipment malfunction, and to check for current test inspection dates. In addition, some companies also perform interchange inspections when the carrier changes (e.g., when the intermodal tank container is delivered via truck at a plant site prior to the plant taking responsibility for the container).

### ■ Tank and Valve Testing

Tank and valve test inspections are based upon both DOT and IMDG regulations. Intermodal tank containers must be hydrostatic tested every five years, with the test witnessed by a certified approval agency (e.g., American Bureau of Shipping, Marine Container Equipment Certification Corporation). At least every 2-1/2 years (30 months), the tank must be visually inspected and the spring-loaded pressure relief valve removed from the tank and tested. Upon successful completion of these tests, the tank data plate must then be marked with the month and year of the test. Tank containers marked as complying with AAR-600 must be marked and stenciled with the test dates and the next test due dates. The next due date for periodic testing will be shown on the data plate as required by UN- and DOT-adopted recommendations.

## Summary

Most intermodal portable tank containers are manufactured to a standard 20-foot length (6.058 m). About 90% of the tanks in circulation are made of stainless steel that is supported and protected by a steel frame. They are used to transport a varied and diverse range of commodities, including an increasing number of hazardous materials.

Intermodal portable containers have several important markings to help identify the dangerous cargo they may carry. They must be marked with owner's marks and numbers that are found on the right-hand side facing the tank and on both ends. The specification marking indicates the standards to which the tank was built. Tanks may also have hazardous materials placards or other markings that convey safety information or special cargo handling instructions.

The two most widely used specification tank containers permitted to transport bulk liquid hazardous materials in North America are the IM 101 and the IM 102. Although classified as "nonpressurized," these containers can have a working pressure up to 100 psig—which can present safety hazards if domes, valves, and attachments are not handled properly.

Both the IM 101 and IM 102 tanks may be equipped with special safety devices, especially when they transport Class-3 flammable liquids. These can include pressure/vacuum relief valves, burst disks, and an emergency remote shutoff device that can be manually or automatically actuated.

## References

1. Brassington, Bill. 2013. *"Safe Handling of Containers" Safety Briefing Pamphlet Series #30.* International Tank Container Organisation. Romford, Essex, United Kingdom.
2. International Tank Container Organisation, 2015 *Tank Container Fleet Survey.* Romford, Essex, United Kingdom.
3. Noll, Gregory G. and Michael S. Hildebrand. 2014. *Hazardous Materials: Managing the Incident,* 4th ed. Burlington, MA: Jones and Bartlett Learning. pp. 155–156.
4. Union Pacific Railroad Company. September 2006. *A General Guide to Tank Containers.* Omaha, Nebraska.
5. United Nations Economic Commission for Europe. February 2013. *Code of Practice for Packing of Cargo Transport Units (CTUs).* Annexes, New York.

# Intermodal Tank Containers for Gases

## Chapter Outline

- Key Terms
- Introduction
- Pressure Tank Containers
- Cryogenic Tank Containers
- Multiple Element Gas Containers
- Summary
- References

## Key Terms

**Cryogenic Tank Containers** Containers that transport refrigerated liquefied gases such as argon, helium, oxygen, nitrogen, and ethylene. They include the UN Portable Tank T75 and IMO Type 7.

**Multiple Element Gas Containers (MEGCs)** Assemblies of UN cylinders, tubes, or bundles of cylinders interconnected by a manifold and assembled within a rigid frame with corner castings for the transport of nonrefrigerated compressed gases. May also be called tube modules.

**Pressure Tank Containers** Containers that transport nonrefrigerated liquefied compressed gases (Class 2) and chemicals under pressure, including nonrefrigerated liquefied compressed gases (UN portable tank T50, DOT Specification 51, IMO Type 5).

## Introduction

In Chapter 5 we provided an introduction to intermodal tank containers and discussed the nonpressure tank containers that transport liquids and solids (Hazard Class 3–9). In this chapter we will discuss:

1. Pressure tank containers that transport nonrefrigerated liquefied compressed gases (pressure tank containers—UN portable tank T50, DOT Specification 51, IMO Type 5);
2. Cryogenic tank containers that transport refrigerated liquefied gases (UN portable tank T75, DOT IMO Type 7); and
3. Multiple Element Gas Containers (MEGCs), assemblies of UN cylinders, tubes, or bundles of cylinders interconnected by a manifold and assembled within a rigid frame with corner castings for the transport of nonrefrigerated compressed gases.

## Pressure Tank Containers

Pressure tank containers that meet UN portable tank T50 specifications are used to ship liquefied compressed gases and chemicals under pressure. These tanks are comparable to MC331 cargo tanks but with a higher Maximum Allowable Working Pressure (MAWP). The MAWP varies depending on the product being shipped in the tank container.

All new pressurized tank containers must be built to UN T50 instructions as described in U.S. DOT 49 CFR 178.276. While still in circulation, DOT specification 51 and IMO Type 5 tank containers are no longer authorized for manufacture, although they can be used to transport hazardous materials as long as they meet the UN regulations.

Pressure tank containers are used to transport nonrefrigerated liquefied compressed gases, such as Liquefied Petroleum Gas (LPG), anhydrous ammonia, high vapor-pressure flammable liquids, pyrophoric liquids such as aluminum alkyls, and other highly regulated materials including those products in UN Packing Group 1. (See **FIGURE 6-1 A-B** .) T50 tanks are usually manufactured out

Courtesy of Bill Hand.

**FIGURE 6-1** Intermodal pressurized tank containers transport liquefied gas products.
Courtesy of Hildebrand and Noll Associates, Inc.

of mild steel, but some are made out of stainless steel for sensitive products.

## Pressure Tank Container Features

Pressure tank containers (UN portable tank T50, DOT Specification 51, and IMO Type 5) have the following general features:

- **Capacities**—The 20-foot T50 tank container ranges in capacity from 4,500 to 6,367 gallons (17,034 to 25,101 L), while the 40-foot beam-constructed unit has a capacity of 10,610 gallons (40,160 L). Some larger capacities may be found in circulation.
- **Dimensions**—The standard dimension of the T50 tank is 20 ft. long × 8 ft. wide × 8 ft., 6 in. high. While 20 ft. in length is the industry standard, it is not uncommon to find pressure tank containers with a variety of dimensions, depending upon customer and shipper requirements. Several T50 pressure tank container manufacturers make custom tanks for special requirements. Specialized pressure containers as small as 50 gallons can be found in some situations (e.g., pyrophoric liquids). (See **FIGURE 6-2 A-D**.)
- **Operating Pressure and Shell Thickness**—T50 pressure tank containers generally have a 0.772-inch (19.6 mm) shell thickness and 0.773-inch (19.6 mm) minimum head-shell thickness. The tanks are equipped with internal stiffening rings for additional tank support.

   The T50 internal operating pressure can vary depending on whether the tank is insulated or equipped with a sunshield. Several T50 tank manufacturers consulted by the authors build their pressure tanks to an MAWP ranging from 261 to 493 psig (18 to 34 bar).

Specification 51 containers that are still in service have internal operating pressures ranging from 100 to 500 psig (6.9 to 34.5 bar).

## Data Plate

Each new UN portable tank T50 must have a corrosion-resistant metal data plate permanently attached to the portable tank in a conspicuous place and readily accessible for inspection (49 CFR 178.274(i)).

   Under 49 CFR 178.274(i) the following information is required on the data plate:

- Country of manufacture
- U N Approval Country
- Approval Number
- Alternative Arrangements (see § 178.274(a)(2)) "AA"
- Manufacturer's name or mark
- Manufacturer's serial number
- Approval Agency (authorized body for the design approval)
- Owner's registration number
- Year of manufacture
- Pressure vessel code to which the shell is designed
- Test pressure in bar gauge
- MAWP in bar gauge
- External design pressure (not required for portable tanks used for refrigerated liquefied gases) in bar gauge
- Design temperature range in °C (For portable tanks used for refrigerated liquefied gases, the minimum design temperature must be marked.)
- Water capacity at 20°C/L
- Water capacity of each compartment at 20°C in liters
- Initial pressure test date and witness identification
- MAWP for heating/cooling system in bar gauge
- Shell material(s) and material standard reference(s)
- Equivalent thickness in reference steel in mm

**FIGURE 6-2** UN portable tank T50, DOT Specification 51, and IMO Type 5 tanks can be found in a variety of sizes and capacities.
Courtesy of Hildebrand and Noll Associates, Inc.

- Lining material (when applicable)
- Date and type of most recent periodic test(s)
- Month in Year_____ Test pressure in bar gauge
- Stamp of approval agency that performed or witnessed the most recent test
- For portable tanks used for refrigerated liquefied gases:
  - Either "thermally insulated" or "vacuum insulated"
  - Effectiveness of the insulation system (heat influx) in watts (W)
  - Reference holding time days or hours and initial pressure in bar/kPa gauge and degree of filling in kg for each refrigerated liquefied gas permitted for transportation

The following information must be marked either on the portable tank itself or on a metal plate firmly secured to the portable tank:

- Name of the operator
- Name of hazardous materials being transported and maximum mean bulk temperature (except for refrigerated liquefied gases, the name and temperature are only required when the maximum mean bulk temperature is higher than 50°C.)
- Maximum permissible gross mass (MPGM) in kg
- Unladen (tare) mass in kg

## ■ Pressure Tank Container Fittings

UN Portable Tanks T50, Spec 51, and IMO Type 5 containers have a number of fittings to make them both safe and functional. These fittings may be on the top, the end, or the bottom side of the tank container. Generally, the fittings are enclosed with a cover or recessed to protect them from mechanical damage.

The following describes the most common fittings found on pressurized tank containers:

- **Loading/Unloading Valves**—The characteristics and nature of liquefied gases and high vapor-pressure products requires that they be loaded and unloaded using different methods than required for nonpressurized liquids. Liquid and vapor valves are used for both filling and emptying the tank. The liquid valve extends into the lading by means of an eduction pipe, which may also be fitted with an excess flow check valve. Vapor valves, which may have an excess flow check valve, are used to remove vapors from the tank or to pressurize the tank for unloading. (See **FIGURE 6-3**.) As a safety feature, the liquid and vapor valve connections are of different sizes.

The UN T50 tank must be equipped with three serially mounted and "independent" shut-off devices for both the liquid and vapor lines. In simple terms, the operator has three different ways to stop the flow of product:

- An internal valve that may be actuated by a foot valve, an excess flow valve, or external cable or rod during transfer operations;
- An external valve; and
- A threaded cap or bolted blank flange.

Pressure tank containers will have a liquid valve and a vapor valve. Each is required to be marked accordingly. Based upon the position of the container in an emergency scenario, either set of valves may be used for loading and unloading.

All tank outlets must be marked to designate vapor or liquid discharge potential when the tank is filled to the maximum level permitted. They may be either threaded or flanged valves. Remember that valve threads may be British Standard Pipe (BSP) or metric threads.

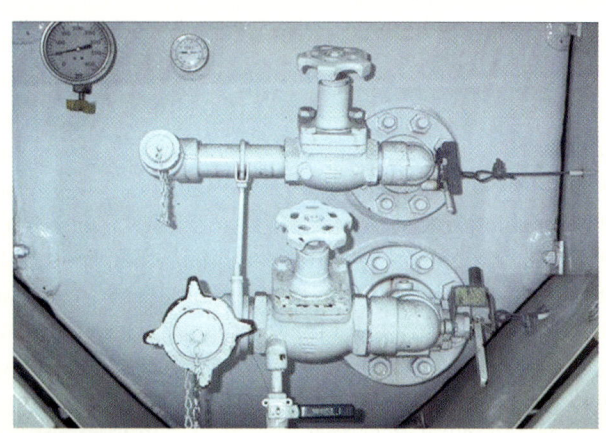

**FIGURE 6-3** Liquid and vapor valves are used for both filling and emptying the tank. *Note:* The vapor line is on top and the liquid line is on the bottom.
Courtesy of Hildebrand and Noll Associates, Inc.

- **Gauging Devices**—May be installed to measure how much liquid is in the tank container. Various types may be found, including rotary gauges and open or closed gauging devices similar to those found on some pressurized railroad tank cars. Open-type gauging devices measure tank outage (i.e., vapor space) by releasing liquid when the liquid level reaches the bottom of a tube. Closed-type gauging devices measure outage with a float and a magnet on a measuring rod or dial indicator.
- **Sample Lines**—Are used for sampling the lading without opening the tank. Sample lines can develop leaks at the handle or around the plugs.
- **Thermometer Wells**—The thermometer wells generally are not being installed on new tanks unless requested by the customer. They are used for measuring the lading temperature. Temperature readings can assist responders in determining if a product is expanding and increasing the internal tank pressure using temperature/pressure curve charts for the lading. If a thermometer well tube breaks inside the tank container, leaks can develop.
- **Thermometer Gauges**—Instruments for measuring and indicating the temperature of the liquid inside the tank. They may read in Fahrenheit, Celsius, or dual scale. The gauge typically measures a range (e.g., −76°F to +212°F (−60°C to +100°C). The gauge is fitted at the rear end of the tank.
- **Manholes**—Are typically bolted and found at the end of the tank container. Unless a repair or inspection is required, they are never unbolted. (See **FIGURE 6-4**.)
- **Sun Shields**—Work like a beach umbrella; they shield the tank's shell from the sun. The sun shield is attached to the top of the tank container. UN T50 tanks are seldom manufactured without a sun shield because unshielded "bare" tanks would require a higher maximum allowable working pressure and are more expensive to manufacture. They also have a higher tare weight, which translates to less cargo capacity and lost revenue. (See **FIGURE 6-5**.)

## ■ Pressure Tank Container Safety Devices

Because the UN Portable Tank T50, DOT Spec. 51, and IMO Type 5 tanks are pressure vessels, they must be equipped with safety devices to reduce the risk of

**FIGURE 6-4** Manholes on DOT Spec 51 tanks are bolted and only opened during maintenance and inspection.

Courtesy of Glen Rudner.

overpressure and tank failure under normal operating conditions. The following safety devices are found on the pressure tank containers:

- **Pressure Relief Devices**—Pressure relief devices, usually a spring-loaded pressure relief valve, are mounted on top of the container in the vapor space to protect the tank from over-pressure under abnormal conditions, such as fire impingement or an internal chemical reaction. If highly corrosive materials like hydrogen fluoride are

transported, the valve working parts are located on the outside of the tank and protected by a rupture disk between the valve assembly and the lading. Remember that pressure tank containers carrying LPG can fail violently if impinged upon by fire, even if the pressure relief devices have actuated (e.g., BLEVE scenario). (See **FIGURE 6-6** .)

- **Excess Flow Valves**—As noted above, excess flow valves may be found on both liquid and vapor piping. Mounted inside the tank under the liquid and vapor valves, excess flow valves will stop the product flow if a valve is sheared off. Excess flow valves are operated by gravity or pressure differential. An important thing to remember about excess flow valves is that there must be an excessive flow for these devices to operate. They are not designed to stop the flow of a minor line break, which could be considerable with high expansion ratio products like anhydrous ammonia that has an expansion ratio of 1 to 850 (i.e., 1 gallon/L of liquid ammonia will expand 850 times to ammonia vapor).

## Cryogenic Tank Containers

The UN Portable Tank T75 and IMO Type 7 cryogenic tank container transports refrigerated liquefied gases such as liquid argon, liquid helium, liquid oxygen (LOX), liquid nitrogen, and ethylene. If local codes

**FIGURE 6-5** UN T50 specification tanks are usually equipped with a sun shield. An opening is made for the pressure relief valve to vent. Note the Do Not Lift marking in the center.

Courtesy of Hildebrand and Noll Associates, Inc.

**FIGURE 6-6** Pressure relief devices are mounted on top of the container in the vapor space to protect the tank from over-pressure.

Courtesy of Hildebrand and Noll Associates, Inc.

allow it, they can also serve as stationary Liquefied Natural Gas (LNG) storage tanks where, for various reasons, building a temporary LNG stationary satellite station is not profitable.

The T75 and IMO Type 7 cryogenic intermodal tanks make up a very small percentage of the worldwide intermodal container fleet. The primary reason is that they do not have any refrigeration equipment; they are basically a big thermos bottle. Intermodal cryogenic containers rely on vacuum-jacketed and thermal insulation to reduce heat gain and keep the cargo in a liquefied state during transit. (See **FIGURE 6-7** .)

## ■ Holding Time

Cryogenic tank containers have a calculated "One Way Holding Time" that determines how long the product can remain in a cryogenic state. One Way Holding Time is defined as the maximum safe time period between the initial filling time to the destination arrival time where ambient heat sources will not cause an internal pressure rise and result in the activation of pressure relief devices. Holding time will vary depending on the tank and product. Product handlers need to know the holding time to ensure the container arrives at its destination on time and in a liquid state. (See **FIGURE 6-8** .)

Cryogenic tank containers are designed, constructed, and operated in accordance with the following:

- Portable tank provision T75 and special provision TP5 of the Hazardous Materials Tables under 49CFR 172.101
- 49 CFR 178.274 and 178.277 for UN portable tanks

**FIGURE 6-7** The DOT Spec. 51 (IMO Type 7) or UN T75 cryogenic tank container transports refrigerated liquefied gases.

Courtesy of Hildebrand and Noll Associates, Inc.

**FIGURE 6-8** Cryogenic tanks are like giant thermos bottles. They have a limited holding time for transit to their destination.

Courtesy of Hildebrand and Noll Associates, Inc.

- 49 CFR 178.338 for cargo tanks
- Special Permit such as DOT-SP-11186

Each tank must be prepared and shipped as required in 49 CFR 173.318 as applicable to the lading. The maximum gross weight of each tank must be as follows:

- Liquid Argon—32,300 pounds
- Liquid Nitrogen—23,600 pounds
- Liquid Oxygen—28,700 pounds

## ■ Features

Cryogenic tank containers have the following general features:

- **Size**—Standard T75 cryogenic tank containers are either 20 ft. or 40 ft. in length.
- **Capacity**—Capacities can range from 2,113 to 14,265 gallons (8,000 to 54,000 L); however, specialty containers may be manufactured in 10-, 30-, and 45-ft. lengths.
- **Design Pressure**—T75 tanks can be manufactured for 150, 250, and 320 psig.
- **Tank Shell**—Cryogenic tank containers will have an outer metal jacket covering the entire pressure vessel with an interstitial space that is a vacuum. It functions just like a thermos bottle and insulates the tank so the product can remain in a liquid state longer. Some cryogenic tanks are equipped with a vacuum jacket that is filled with liquid nitrogen.
- **Fittings**—Fittings for the T75 cryogenic tank are complex. There are numerous valve and fitting options offered by tank manufacturers to meet customer loading and offloading requirements based on the type of service required and the characteristics of the product. It is not unusual

to see ice formed up on fittings. This is sometimes the source of emergency calls when someone thinks the product is leaking from a fitting.

The tank may be equipped with one or more valve compartments. Each compartment will be equipped with a valve for handling liquid and vapor. There will be three valve closure methods. This normally includes two external valve closure options and an inner valve. The inner valve is usually pneumatically operated and requires a source of pressure to be activated.

- **Cryogenic Pumps**—Some cryogenic tank containers are equipped with a cryogenic pump, usually located in a compartment on the side of the container. To be placed in service, they need to be cooled down to the temperature of the liquid inside the tank. Most cryogenic liquids are stored around −180°F (−118°C).
- **Pressure Control**—Cryogenic tank containers are manufactured for a given pressure rating based on the tanks "Holding Time." While the tank is designed to minimize internal heating of the product during transit, ambient heating does occur and some of the cryogenic liquid will pass from a liquid state and boil off into a vapor state. This generates internal pressure that is well within the tanks Maximum Allowable Working Pressure (MAWP) design parameters, but must be safely relieved.

Each cryogenic tank container is required to be equipped with one or more pressure control devices. Each pressure relief valve and frangible disk must be arranged so that the discharge will not impinge upon any of the vessel, the tank itself, or any other cargo. T75 cryogenic tank containers are equipped with a bleed valve specially designed for the container based on the type of product being transported. It can be switched on and off manually to relieve pressure. On a warm day you can sometimes see cold vapor being vented on tanks being transported by truck. The bleed valve is in the closed position when on board ship. Cryogenic tanks are always stowed on deck and usually in the stern of the vessel.

In addition to the features described above, the tanks are equipped with a variety of equipment that allow the operator to monitor, load, and unload the cargo. These can include:

- Controls for filling levels
- Internal coils for raising the pressure for unloading the tank
- Flow gauges
- Vacuum connection for checking or drawing a vacuum

## Multiple Element Gas Containers (MEGCs)

Multiple Element Gas Containers (MEGCs) are assemblies of UN cylinders, tubes, or bundles of cylinders interconnected by a manifold and assembled within a rigid frame with corner castings for the transport of nonrefrigerated compressed gases. MEGCs are also popularly known in the United States as tube modules.

Examples of nonrefrigerated compressed gases shipped in MECGs include oxygen, nitrogen, helium, methane, hydrogen, chlorine, and anhydrous hydrogen chloride.

Each pressure receptacle of an MEGC must be of the same design type, seamless steel, and constructed to meet ISO standards. The pressure receptacles are manufactured from stainless steel or mild steel and are engineered to handle high pressure ranging 3,000 to 5,000 psi (20,700 to 34,500 kPa). Each pressure receptacle must be equipped with one or more pressure relief devices. (See **FIGURE 6-9 A-C**.)

MEGC capacities will vary based on the product being transported and size of the pressure receptacle. Seamless pressure receptacles mounted horizontally can range from 9 in. to 48 in. in diameter (22.86 cm to 121.92 cm). The standard MEGC module length is 20 feet, but they can also be found in 40-foot lengths or smaller sizes for specialty products. These high-pressure cylinders are built to DOT and other international specifications.

MEGC pressure receptacles equipped with pressure relief devices must function at not less than 70% of cylinder test pressure. Valves are located in a cabinet at the end of the cylinder frame. If an MEGC element is intended to transport toxic gases, each element must be equipped with an isolation valve. Anhydrous hydrogen chloride cylinder tubes are equipped with a dip tube so that the product can be used either as a gas or liquid. (See **FIGURE 6-10**.)

MEGCs are required to have a corrosion-resistant data plate attached to the container.

## Summary

Intermodal tank containers that transport Hazard Class 2 gases include those that transport liquefied gases, cryogenic liquids, and high vapor-pressure products. Old DOT specifications are valid for intermodal pressurized

Courtesy of Hildebrand and Noll Associates, Inc.

Courtesy of Hildebrand and Noll Associates, Inc.

Courtesy of Glen Rudner.

**FIGURE 6-9** MEGCs are simply large cylinders mounted horizontally inside a box or steel frame.

tank containers still in service as long as they conform to the new United Nations recommendations. However, they are not authorized for new construction.

Pressurized tank containers are used to transport liquefied gases, such as Liquefied Petroleum Gas (LPG), anhydrous ammonia, high vapor-pressure flammable liquids, pyrophoric liquids such as aluminum alkyls, and other highly regulated materials. Because the Specification 51 tank is a pressure vessel, it must be equipped with safety devices to reduce the risk of overpressure and tank failure under normal operating conditions. Tanks are equipped with a pressure relief device and excess flow valves.

The UN T75 cryogenic tank container (DOT Spec. 51 or IMO Type 7) transports refrigerated liquefied gases, such as liquid argon, liquid carbon dioxide, liquid helium, liquid oxygen (LOX), liquid nitrogen, and ethylene. They do not have any refrigeration equipment; they rely on vacuum-jacketed and thermal insulation to reduce heat gain and keep the cargo in a liquefied state during transit.

MEGCs are assemblies of UN cylinders, tubes, or bundles of cylinders interconnected by a manifold and assembled within a rigid frame with corner castings for the transport of nonrefrigerated compressed gases. Also known as tube modules, they are used to transport nonrefrigerated oxygen, nitrogen, helium, hydrogen, and anhydrous hydrogen chloride.

**FIGURE 6-10** Cylinder valves are located at the end of the tank.

Courtesy of Mike Hildebrand.

## References

1. Brassington, Bill. 2013. *"Safe Handling of Containers" Safety Briefing Pamphlet Series #30.* Romford, Essex, United Kingdom: International Tank Container Organisation. pp. 17–20.
2. United Nations. 2015. Recommendations on the Transport of Dangerous Goods - Model Regulations, Volume I – Part 6.7. *Requirements for the Design, Construction, Inspection and Testing of Portable Tanks and Multiple Element Gas Containers (MEGCs).* 19th ed. New York.

# CHAPTER 7

# Intermodal Container Operations

## Chapter Outline

- Key Terms
- Introduction
- Highway Operations
- Railroad Operations
- Marine Operations
- Fixed Facility Operations
- Summary
- References

## Key Terms

**Container On A Flat Car** (COFC) An intermodal container shipped on a railroad flat car.

**Containership** A cargo vessel designed and constructed to transport, within specifically designed cells, intermodal portable tanks and freight containers which are lifted on and off with their contents intact.

**Trailer On A Flat Car** (TOFC) Truck trailers shipped on a railroad flat car.

## Introduction

The intermodal transportation network is vast and joins two or more modes of transportation. These primarily involve highway, railroad, and marine modes of transportation. Currently there is limited use of air transportation for intermodal containers, but the increasing size of cargo aircraft may result in more air transport of specialized containers.

According to *The Geography of Transport Systems* by Dr. Jean-Paul Rodrique and Dr. Brian Slack, the decision to transport a commodity by intermodal container is based on five considerations. These include:

- **Nature and Quantity of the Cargo**—Intermodal containers are ideal for finished goods in loads of less than 25 tons. Whatever is being shipped must fit into standardized containers.
- **Sequence of Transportation Modes**—Known in the shipping industry as the "intermodal transport chain." Rail, highway, barges, and ships make up the chain. The mode of transportation to be used and the sequence of use determine how cost effective it will be to use intermodal containers. The first and last miles on the journey are usually by highway.
- **Origins and Destinations**—Travel distances greater than 310 miles (500 km) usually require intermodal transportation. The reason for this is that 310 miles is an average trucking day, which is approximately 6 to 8 hours.
- **Value of the Cargo**—Intermediate value cargos are well suited for intermodal transportation. High-value cargo typically ships by air while low-value cargo is transported by rail, highway, or ship.
- **Frequency of Shipments**—If the cargo flow needs to be continuous and in similar quantities, intermodal shipping is more efficient and cost effective. For example, if a manufacturer requires a continuous flow of precursor chemicals to make a finished product at a chemical plant, portable tank containers can provide an uninterrupted supply chain.

This chapter will provide an overview of the three primary intermodal container operations including highway, rail, and marine. It will also outline how intermodal containers are used in fixed facilities.

## Highway Operations

Highway movement of intermodal containers falls into two general categories: (1) interterminal transfer operations between different modes of transportation, and (2) the movement of the container from the

**FIGURE 7-1** Movement of intermodal container by highway from the transfer interchange point to the point of final destination.
Courtesy of Bill Hand.

transportation interchange point to its final destination. (See **FIGURE 7-1**.) The handling equipment used for interterminal operations will be discussed under Marine Terminal Operations; this section will focus on highway operations from the transportation interchange point to the final destination.

## ■ Highway Equipment

Several types of chassis can be used to transport intermodal containers. Chassis lengths can vary from 20-feet (6 m) chassis to 48 feet (48 m).

Containers are attached to the chassis by four twist-lock securing devices (i.e., corner castings) located on the four corners of both the chassis and the container. (See **FIGURE 7-2**.) Any other method of attaching the container to the chassis is not permitted.

Drop deck or U-deck truck chassis with operable twist-lock securing devices are used when transporting

**FIGURE 7-2** Standardized corner castings allow intermodal containers to be secured during highway transportation.
Courtesy of Bill Hand.

portable tank containers. They provide a lower center of gravity, and experience shows that they are safer to use for loaded intermodal tank containers. Loaded portable tank containers may also be placed on 28- to 48-feet straight or slider chassis. The entire vehicle must also meet all applicable axle weight and gross vehicle weight laws. (See **SCAN SHEET 7-A**.)

One common challenge for shippers transitioning their intermodal container to the truck chassis is for the container to be legally compliant with gross vehicle weight (GVW) and the distribution of the containers' weight across the vehicle. The U.S. Federal Gross Vehicle Weight Limit is 80,000 pounds (36,287 kg) for the tractor, chassis, and container. For rail, the 53-foot Container On A Flat Car (COFC) domestic intermodal container is about 2,500 pounds heavier than a standard dry van. This works out to a maximum bill of lading weight not over 42,500 pounds (1,9277 kg) for an intermodal load versus the 45,000 pounds shippers are accustomed to for a truckload shipment. The 42,500 pound recommendation is based on the average container, chassis weight, and tractor weight. The shipment must be legal at both the point of origin and destination ramps. The 42,500 pounds weight is well under the 80,000 pound maximum, but many highway shipments leave the container terminal with a day cab tractor and half a fuel load and then transition to a long haul tractor equipped with larger engines, more fuel load, and a sleeping cab. This additional weight can push the vehicle and cargo weight over the 80,000 pound limit.

The weight of intermodal freight containers being transported via highway can be a major safety and operational issue. Freight charges for intermodal freight containers are typically based upon a charge per box without regard to weight. Since weight limits for freight containers are much higher than U.S. highway weight limits, there is a potential for overweight freight containers to be involved in a highway accident (e.g., rollovers, especially on highway ramps). This higher weight limit results in a decreased vehicle braking capacity, over-stressed tires, and a potentially higher center of gravity, all of which negatively impair the driver's control of the vehicle.

## Railroad Operations

According to the Federal Railroad Administration (FRA), the U.S. freight rail network is widely considered one of the most dynamic systems in the world. The $60 billion industry consists of 140,000 rail miles (225,308 km) operated by seven different Class-I railroads with operating revenues of $434.2 million.

Railroads have an intermodal container logistical advantage over highway transportation because they can haul large quantities of freight over long distances with fewer weight restrictions. Railroads also do not have to deal with highway traffic and congestion and are generally reliable to deliver cargo on time. While derailments sometime occur, trains are less prone to accidents than trucks moving over the highway.

Railroads form the intermodal container land bridge in North America between Europe and Asia. The land bridge concept has been very successful at servicing national and international markets and will continue to be a key factor in international shipping. (See **FIGURE 7-3** .)

### ■ Equipment

The traditional lines between the trucking and railroad industries have become seamless to the point where today, virtually every major railroad operates some form of "piggyback service" that interfaces with the trucking industry.

**FIGURE 7-3** Railroads form the land bridge for the intermodal container system.
Courtesy of Mike Yuhas.

Container On A Flat Car and Trailer On A Flat Car (TOFC) are the two most common railroad shipments found today in North America. Both COFCs and TOFCs are used to transport the intermodal freight container. (See **FIGURE 7-4** .)

TOFC operations involve loading a semi-trailer chassis with an intermodal freight container directly on to a railroad flat car. COFC operations require a stacker or a crane to pick up the container in the storage yard and load the intermodal freight container directly onto the flat car. (See **FIGURE 7-5** .)

There are no FRA restrictions specifying what regulated materials may be shipped in COFCs and TOFCs, although there may be individual railroad restrictions. Intermodal portable tank containers cannot be shipped as a TOFC on a flat car except as authorized by 49 CFR 174.63.

### ■ COFCs

COFC operations are simple: a truck delivers the container to a rail terminal and a crane lifts the container off the truck's chassis and places it on a flat car. The process is reversed at the destination point.

COFCs can be found in various configurations on trains, including single and double stacks. The containers are attached to the flat car through twist-lock attachments (i.e., corner castings), similar to those used with truck chassis. The latest generation of deep-well flat cars allows up to four 20- or two 40-foot containers to be double-stacked. Stacking cranes and container lifters are the most common methods of placing COFCs onto and removing them from trains at rail terminals.

COFCs can accommodate specialized containers, and many specialized shipments require specialized COFCs to handle the cargo. (See **FIGURE 7-6** .)

### ■ TOFCs

In the early days, prior to the TOFC concept, trucks would drive to a rail terminal and the cargo would be unloaded from the truck, loaded into box cars, then unloaded at the destination and reloaded on another truck—a very slow and laborious process.

When TOFC operations first started, a semi-trailer was driven into a rail yard and backed onto a flat car and the tractor was disconnected. When the train reached its destination, another truck was connected to the trailer and driven off of the flat car. This was an improvement over box cars, but still required time.

Today, trailers are normally loaded on a flat car by a stacking crane. A spreader is attached to the top corners of the container and the trailer is lowered onto the flat-car. Container forklifts and container lifters can also be

Courtesy of Bill Hand.

© Mike Yuhas. Used with Permission.

**FIGURE 7-4** (A) Container On A Flat Car. (B) Trailer On A Flat Car.

used to lift a trailer onto a flatcar; however, not all trailers are designed to be picked up in this manner and there are specific lifting points on the trailers. Although not as common, ramps (i.e., "circus ramps") may also be found by which the trailer is driven onto the flatcar. The trailer is attached to the flat car by locking the trailer's kingpin onto a fifth wheel assembly built onto the flatcar.

## Intermodal Portable Tank Containers

The transportation of portable tank containers is regulated by 49 CFR 174.63. Key FRA regulatory requirements and Association of American Railroad (AAR) rules

for the movement and switching of portable tank containers include the following:

The transportation of portable tank containers is regulated by 49 CFR 174.63. Key FRA regulatory requirements and AAR rules for the movement and switching of portable tank containers include the following:

- Only AAR 600 intermodal portable tank containers are approved for the transportation of hazardous materials in container-on-a-flat car (COFC) service.
- Flat cars carrying intermodal portable tank containers shall not be cut off while in motion, and may not be kicked or humped.

**FIGURE 7-5** (A) Stackers bring the container from the container yard to the lading area. (B) Cranes span the container lift vertically off the ground and then set it on the COFC where it is locked down.

Courtesy of Bill Hand.

**FIGURE 7-6** Most COFCs are set up to handle 20- and 40-foot containers, but they may also be designed for special freight. This COFC is hauling a fan blade for a commercial windmill power generator.

Courtesy of Bill Hand.

**FIGURE 7-7** Marine terminal operators provide wharfage, dock, warehouse, or other marine facilities to ocean common carriers moving cargo in ocean-borne commerce.

© trekandshoot/Shutterstock.

- Cars moving under their own momentum may not strike any flat car carrying portable tank containers.
- Cars carrying portable tank containers must not be placed next to cars placarded "EXPLOSIVES 1.1 and 1.2," "POISON 6.1, Packing Group 1" (Poison-Inhalation Hazard), "POISON GAS 2.3," or "RADIOACTIVE MATERIALS" (Yellow III label).

## Marine Operations

The United States is a maritime nation with a well-developed water-based transportation system for moving heavy cargo. The U.S. maritime shipping capability is far reaching and includes major deep water ports along the Pacific, Atlantic, and Gulf Coasts, as well as the Mississippi River System, the Saint Lawrence Seaway, and the Great Lakes. If the hundreds of smaller rivers and canals are included, there are very few locations in the continental United States that are not affected by some type of water-based intermodal container operations. (See **FIGURE 7-7**.)

Ninety percent of the world's cargo moves by intermodal container ships. According to 2014 data from the World Shipping Council, four U.S. container terminals ranked among the top 50 most active container terminals in the world. Los Angeles ranked worldwide as #19, handling 8.3 million containers; followed by #21-Long Beach (6.82 million); #26-New York (5.77 million); and #44-Seattle (3.43 million). These four U.S. ports handled about 24 million containers in one year. This data doesn't include other busy container terminals like Houston, Jacksonville, Baltimore, and any of the smaller inland and near-shore ports.

The big emphasis in the container ship business is speed and container capacity. Container ships are larger and faster than general freighters. The Hawaiian Enterprise class, for example, can transport over 10,000 containers between San Francisco and Hawaii in three and one half days. The Evergreen Line's 16 G-Type ships can circle the globe in 80 days. The newest container ship *Benjamin Franklin* went in service in 2015 and can carry 18,000 TEUs! The typical container ship has a crew of about 15 personnel. In the harbor, the turnaround time is often less than a single eight-hour shift depending on the number of TEUs.

Despite the wide range of options, transportation by water is still the most cost-effective method for moving bulk cargo. Several important factors account for the increasing number of bulk intermodal shipments via water. These include:

- **Advances in Crane Technology**—A modern container terminal will be equipped with one or more giant ship-shore gantry cranes that can reach out over the ships, sometimes an amazing 25 containers wide. These cranes are rail mounted and able to traverse up and down the pier to exactly align with the bays over the ship's holds. The outreach of the crane's jib permits a carriage to roll from over the quay to over the ship, a spreader with four locks making fast to the corners of the container and releasing them when grounded. The most up-to-date cranes can handle two boxes simultaneously in an increasingly automated process. What once took a hundred people two days can now be offloaded using ten people in several hours. The largest container ships can be turned around in less than 24 hours.

- **Information Management Systems**—A modern terminal is heavily dependent on computer assistance for all its planning, with the objective of moving the container around as little as possible and delivering it to the gantry crane in the exact sequence to mesh with the ship's cargo plan. This planning process ensures that the containers for one particular port are not buried by those for a later destination. The ship's cargo loading plan also takes into account the vessel's stability at all stages of the voyage and ensures that the structure of the ship is never placed under undue structural stress. The verification of container weights is an important responsibility for container terminals.

  Using computers, bar code technology, and advanced software packages, longshoremen can easily locate and track containers on board vessels. The computer and its software drives the crane, identifies the container to be retrieved, finds its location onboard the ship, sorts and prioritizes the sequence for offloading, removes it from the ship, and transfers it to the correct staging area on the dock, or to a waiting truck or rail car. All of these tasks are accomplished in one fluid motion in a matter of minutes.

- **Upgraded Ports and Marine Terminals**—The decline of the general cargo and break bulk industry, coupled with the rapid transition to intermodal containers, left many U.S. ports playing catch-up to the more modern facilities constructed in Europe and Asia. U.S. ports have upgraded their facilities specifically for handling intermodal containers. Other major improvements include a better trained workforce and improved access of railways and highways to marine terminals. Improvements to the railroad and highway infrastructure near major ports have made intermodal transportation more cost effective for inland shipments.

- **Removal of Trade Barriers and Adoption of International Standards**—Globalization, establishment of Free Trade Zones, removal of restrictive tariffs, and more flexible trade agreements have resulted in a wider range of commodities moving back and forth between different countries. Small businesses located inland in North America and Asia have found new markets abroad which were previously not accessible because of trade restrictions or the inability to inexpensively move goods to a major port. The cost effectiveness of intermodal containers combined with a more even playing field in the international marketplace have increased the number of containers in transit. The U.S. adoption of the United Nations (UN) Recommendations on the Transport of Dangerous Goods-Model Regulations has standardized container standards, placards, markings, etc.

## ■ Types of Intermodal Container Ships

Globally, the maritime shipping industry has made a dramatic transition from the break bulk and general cargo carriers to a more modern fleet of container ships. Break bulk carriers are designed to carry cargo in smaller parcels, such as crates, bags, or barrels. In addition to their cargo, they will also carry "dunnage" which is used to support and separate cargo during transit. General cargo vessels can carry a variety of cargo, ranging from large, bulky machinery to smaller, odd-shaped packages. These ships may also have large refrigerated spaces for perishable foods.

Both break bulk and general cargo carriers require large crews of longshoremen to break out the ship's cargo and move it to the dock. Removal of the ship's dunnage and individual small packages is hazardous and time consuming. Although these vessels are still widely used internationally, they are being replaced by container ships.

Container ships are specialized carriers designed to handle intermodal freight containers and portable tank containers. Container ships are usually loaded and offloaded with a gantry crane. Loading and discharge rates are figured by the number of lifts per hour, with an average rate of 20 lifts per hour per crane. (See **FIGURE 7-8**.)

The first-generation container ship (1956–1970) carried 500 to 800 TEUs. The current sixth-generation container ships (2006 to present) can carry 11,000 to 14,000 TEUs. There are many generations of container ships still in service around the world with varying cargo capacities.

There are various types of container ships and they can be categorized in several ways. These include:

- **Generation**—By years built, length, draft, and TEU capacity (e.g., generations 1 through 6). (See **FIGURE 7-9**.)
- **Handling Mode**—There are several configurations of container ships, ranging from customized single-purpose vessels equipped for carrying containers in all available spaces (above and below deck) to the "convertible" or "multiple cargo" container type where some of the cargo space can be used to carry containers while other space is

Courtesy of Bill Hand.

Courtesy of Glen Rudner.

**FIGURE 7-8** (A) This younger generation container ship carries fewer containers. A tug is maneuvering the ship to the pier. Note that there are no cranes mounted on the ship. (B) Container ship tied to the wharf. TEU's are stacked on the stern deck.

reserved for more traditional break bulk or dry bulk cargo or heavy machinery.

Most container ships require a gantry or other type of crane to load and unload the containers at a pier. Most container ships do not have cranes on board to handle cargo, however some do. These are usually smaller ships that must navigate shallow water and reach remote areas. There are also container ships designed for trucks to drive directly on and off the ship via a ramp that lowers to the pier. These ships are known as Roll-On-Roll-Off or RoRo's. (See **FIGURE 7-10** .)

- **Ship Size**—Size does not matter when the container ship is in open deep-water ocean, but it does matter when the ship must pass through canals like the Panama, Suez, and others. Height, width, draft, and length are the key factors. These ships have various designations based on if they were built prior to or after the canals and channels had been improved (e.g., Panamax, Post-Panamax, Suezmax, Post-Suezmax). The Panama Canal Expansion project doubled the capacity of the Panama Canal by adding a new lane. The project took 10 years and cost $5 billion. (See **FIGURE 7-11** .)

**FIGURE 7-9** (A) This container ship is equipped with its own cranes so that it can load and can unload cargo in smaller ports inaccessible to larger vessels due to shallow draft. (B) This container ship is designed to be loaded and unloaded with dockside container cranes. This the most common type of container ship.

Courtesy of Bill Hand.

**FIGURE 7-10** This container ship is transiting the Panama Canal. Some container ships are so large they cannot transit the canal locks.

Courtesy of Dr. William Fish.

- **Specialization**—These are purpose-built container ships to handle specialized cargo or to support special missions. (See **FIGURE 7-12**.)
- **Service Range**—Some smaller container ships or barges are designed as feeder ships that move cargo from its origin (e.g., manufacturing facility or shallow draft river port) to mother ships that are loaded at a major port.

### Container Ship Cargo Handling and Stowage

Container ships can carry intermodal containers both in below-deck cargo holds and stacked above deck. When containers are loaded on deck they are positioned over a deck socket that is equipped with a twist-lock device.

When the container is in position, the crane operator lowers the container in position and all four contact points are locked down manually by a stevedore. This process is the same on highway and rail operations.

Containers are stacked and then lashed together by a stevedore by hooking one end of a turnbuckle to a lashing plate and the other end to a hole in the corner of the container. Once the turnbuckle is in pace, it is tightened down to secure the container. (See **FIGURE 7-13**.)

Once all containers above deck are removed, the hatch covers are removed using a gantry crane. The hatch covers are moved onto the pier until the ship is reloaded with cargo and placed back over the cargo hold. Containers are then removed from below deck. (See **FIGURE 7-14**.)

Within the container yard, a variety of specialized equipment is used to move containers to or from the ship where the gantry crane loads the ship. These include container handlers (see **FIGURE 7-5 A**), container forklifts, and straddle carriers. (See **FIGURE 7-15**.)

## Fixed Facility Operations

Intermodal containers may also be found at fixed industrial facilities. They may be in staging for use, being loaded or offloaded, or temporarily connected to a process unit.

### Temporary Storage

Portable tank containers are commonly used as temporary bulk storage containers at fixed facilities. Examples include the use of glycol solutions at airports for aircraft deicing operations, the use of solvents and adhesives at

**FIGURE 7-11** (A) This specialty container ship is a railroad ship. Intermodal containers are loaded on to TOFCs or COFCs and a yard engine pushes the coupled cars onto the ship, much like a RoRo for cars and trucks. This ship services New Zealand. (B) An intermodal portable tank container loaded onto a COFC debarking the railroad ship.

Courtesy of Mike Hildebrand.

**FIGURE 7-12** (A) Lashing rods are locked in place diagonally from the bottom up and span two containers. This is done by stevedores who do it the old fashion way—hard work. (B) The lashing rod is hooked into a lashing plate and then the turnbuckle is tightened down to keep the container secure. If these fail during a storm at sea, the container can tumble overboard.

Courtesy of Bill Hand.

highway and bridge construction sites, and the use of corrosive liquids, oxidizers, and poisons as intermediates for chemical processing operations.

Among the greatest potential emergency response problems are building and process construction operations. Experience shows that since construction projects are temporary and usually unoccupied by the general public, they may not receive the same level of code inspection and enforcement that a fixed facility would receive (e.g., tank farm or chemical storage area). Consequently, portable tank containers often "show up" on the job site

and remain for the duration without meeting model fire code provisions for separation and diking.

## ■ Permanent and Semipermanent Storage

There is growing interest in using portable tank containers for fixed-facility permanent storage. In this scenario, the owner intends to incorporate the tank container as a permanent part of the facility's storage or process area. The advantage of this arrangement is that the owner can lease the tank container or obtain the container as part of a package deal with

**FIGURE 7-13** (A) Hatches on the deck of a container ship are removed by a gantry crane and lowered to the pier while the cargo hold is loaded or unloaded. Note that there are other hatches shown still in place and each hatch is numbered. (B) With the hatch removed the cargo hold can be unloaded. Note the Cell Guides used to align containers and guide the containers in or out of the cargo hold. Containers below deck are not lashed in place like they are on deck.

Courtesy of Bill Hand.

**FIGURE 7-14** Straddle carriers are used for moving containers around the container terminal or to and from a ship. They come in different sizes and lifting capacities. A longshoreman straddles the container with the carrier and lifts and lowers it.
Courtesy of Bill Hand.

the supplier (i.e., you buy the product from us and we'll supply the tank and install it). If the tank is damaged or requires cleaning, the facility calls the tank owner for a replacement.

In some cases, tank containers may be incorporated into the process operation. If the tank container can be directly tied into the process unit, it eliminates the time, effort, and risks associated with product transfer operations. This arrangement is especially popular among research and development laboratories where several chemical storage tanks may be required during the life of a special project. When the project is completed, the process is shut down and the tank container is removed.

## Loading and Off-Loading Methods

Loading and offloading methods for portable tank containers will vary depending upon the type of product and container (i.e., liquid vs. gas) and the nature of the hazardous material. In general, "open systems" may be used for high flash point, low vapor pressure materials, while "closed systems" will be used for hazardous materials with low flash points and high vapor pressures. If a tank container has been breached and cannot be repaired, product transfer or offloading some or all of the product into another compatible tank container may be required. In some situations, it may be possible to immediately use the product as part of a facility's process operation. Emergency responders should be familiar with the loading and offloading techniques for various types of intermodal tanks.

## Loading Methods

Methods of product loading include the following:

- **Gravity Loading Through the Manhole**—The product is allowed to free-flow by gravity from a storage vessel through the manhole into the tank container. This is an open system and any vapors will be allowed to escape into the atmosphere. A loading arm is normally used in these operations and the dip tube extends to the bottom of the container to help eliminate splash loading and the buildup of static charges.

- **Closed-System Gravity Loading Through Top or Bottom Outlets**—The product is allowed to free-flow by gravity from a storage vessel through the top or bottom valves into the tank container, while vapors are vented back to the storage tank via the air line connection. The top loading valve usually has a dip tube attached that extends to and contacts the bottom of the tank. Sometimes this dip tube is removed while the container is in transit to prevent damage to the inside of the container.

## Off-Loading Methods

Methods of product offloading include the following:

- **Gravity Discharge**—The product is allowed to free-flow by gravity from the bottom outlet to low-level or below-grade storage tanks. Venting must be provided by pressure/vacuum vents to prevent vacuum damage to the tank container.

- **Pressure Discharge**—The product is pumped through the top or bottom valves when pressure is applied to the tank container. Materials transported under an inert gas blanket are normally pumped off using nitrogen or another inert gas as the medium. Under no circumstances should the maximum working pressure of the tank container be exceeded.

- **Pumped Discharge**—The product is pumped from the tank container from hoses connected to the top or bottom valves, or to a hose with a pipe (stinger) dropped through the manhole. Depending upon the nature of the transfer, air or gas must be allowed to replace the cargo being pumped off. Depending upon the product, this can be achieved by either opening the manhole or air line connection, or connecting to an inert gas supply.

  When using a high-capacity pump, it is recommended that a vacuum safety valve be incorporated in the suction hose to prevent the tank container from a vacuum collapse. Depending on the properties of the product being transferred, the air line connection can also be opened to prevent pulling a vacuum on the tank.

- **Pressure Loading Through Top or Bottom Outlets**—The product is loaded by pressurizing the storage vessel and moving the product through the top or bottom valves into the tank container. For sensitive or hazardous materials, the vapors are vented back to the storage tank, a vent tank or a scrubber via the air line connection.

## ■ Safety Guidelines for Transfer Operations

The following basic safety procedures should be followed when performing product transfer operations:

1. All equipment from the facility storage tanks to the tank container, including valves, hoses, pumps, gauges, connections, vapor return lines, etc., should be regarded as part of the "total system" which is exposed to the hazards and properties of the cargo (e.g., viscosity, corrosivity, temperature, pressure). All items should be thoroughly checked for suitability and condition.

2. Ensure that bonding and grounding connections are in place, as necessary.

3. Check that the hose connections have the same thread or fitting as the tank connections. Remember that intermodal containers typically use British Standard Pipe (BSP) or metric threads! Ensure that the correct joint rings and gaskets are used and that they are compatible with the cargo.

4. Check the appropriate facilities and/or procedures for the draining of hoses and valves at the completion of the transfer operation.

5. If the container is being loaded, also perform these additional checks:
   - Depending upon the previous cargo, check if a Cleanliness Certificate or a Gas Free Certificate is required.
   - Open the manhole and examine the tank and discharge valves for cleanliness.
   - Check the pressure/vacuum relief valves for freedom of movement.

6. There will always be a pressure differential between a closed tank and the atmosphere. Valves must be opened slowly. Always relieve the pressure before opening the manhole. Do not stand on the manhole when loosening the wing nuts.

7. When using the bottom outlet, always open or close the foot valve first.

8. Manhole wing nuts only need to be hand-tight. If the gasket is in serviceable condition, further tightening will not be necessary. Always use a star pattern when tightening manhole cover wing nuts (swing nuts).

9. Look around the surrounding area. If there is a leak, determine where will the product go and what the response should be.

10. Never enter a tank until all confined-space safety procedures have been met and proper confined-space permits issued.

# Scan 7-A—Types of Truck Chassis Used for Transporting Intermodal Portable Tank Containers

There are a variety of different chassis configurations available that provide flexibility for the carrier to transport combinations of full and empty portable tank containers. The illustration below provides some examples of chassis options.

- Chassis #1 through 5 can be used to transport one empty tank container.
- Chassis #5 can be used to transport two empty tank containers.
- Chassis # 3, 4, and 5 can be used to transport both empty tank containers and full tank containers. These chassis provide a lower center of gravity and are safer to use for loaded tank containers

**DIFFERENT TYPES OF CHASSIS**

- Chassis 1 through 5 can be used to transport one empty tank container.
- Chassis 5 can be used to transport two empty tank containers.
- Chassis 3, 4 and 5 can be used to transport both empty tank containers and full tank containers. These chassis provide a lower center of gravity and are safer to use for loaded tank containers.

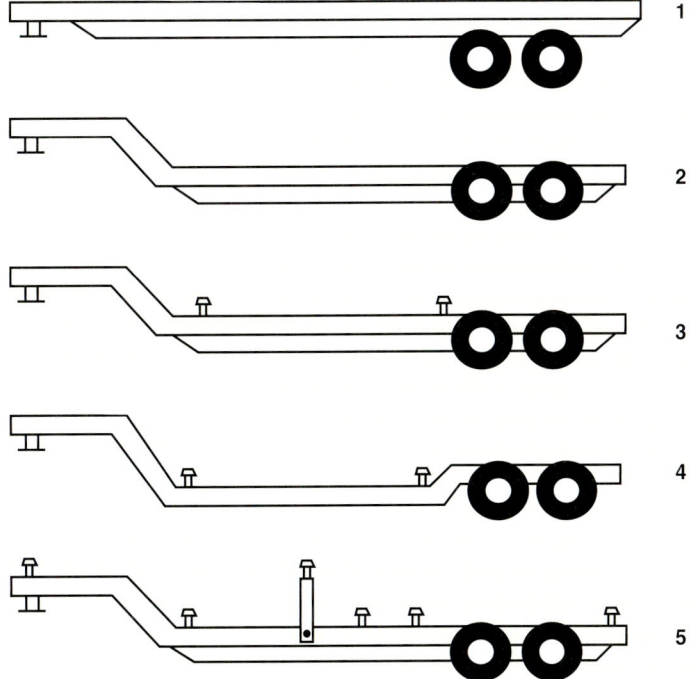

## Summary

The intermodal transportation network is vast and joins two or more modes of transportation. These primarily involve highway, railroad, and marine modes of transportation.

Highway movement of intermodal containers falls into two general categories: (1) interterminal transfer operations between different modes of transportation, and (2) the movement of the container from the transportation interchange point to its final destination.

Railroads form the intermodal container land bridge in North America between Europe and Asia. The land bridge concept has been very successful at servicing national and international markets and will continue to be a key factor in international shipping.

Railroads have an intermodal container logistical advantage over highway transportation because they can haul large quantities of freight over long distances with fewer weigh restrictions.

Today, 90% of the world's cargo moves by intermodal container ships. Container ships are specialized carriers designed to handle intermodal freight containers and portable tank containers. The U.S. maritime shipping capability is far reaching and includes major deep water ports along the Pacific, Atlantic, and Gulf Coasts, as well as the Mississippi River System, the Saint Lawrence Seaway, and the Great Lakes.

Intermodal containers may be found at fixed industrial facilities. They may be in staging for use, being loaded or offloaded, or temporarily connected to a process unit.

## References

1. LaGore, Rick. 2015. "Intermodal Weight – The Most Common Issue for Shippers." IDS: Delivering Competitive Advantage. (September 13).
2. Rodrigue, John-Paul and Brian Slack. 2013. *The Geography of Transport Systems*. 3rd ed. New York: Routlege.
3. Transportation Research Board of the National Academies. 2006. "The Intermodal Container Era: History, Security, and Trends." TR News, No. 246.
4. U.S. Department of Transportation.49 CFR 174.63, Portable Tanks, IM Portable Tanks, IBCS, Large Packagings, Cargo Tanks, and Multi-Unit Tank Cars. Washington, D.C.

# Emergency Response

## Chapter Outline

- Key Terms
- Introduction
- Intermodal Portable Tank Container Emergencies
- Intermodal Freight Container Emergencies
- Industrial Facility Emergencies
- Container Terminal Emergencies
- Highway Transportation Emergencies
- Railroad Transportation Emergencies
- Marine Transportation Emergencies
- Summary
- References

## Key Terms

**Damage Assessment**  The process of gathering and evaluating container damage as a result of a hazardous materials incident.

**Emergency Response**  Response to any occurrence which has caused or could result in the release of a hazardous substance.

**Leak Control**  Actions taken to contain or keep a material within its container.

## Introduction

Intermodal containers can be found at fixed facilities and in all modes of transportation. They can be found just about anywhere. (See **FIGURE 8-1**.) Consequently, both intermodal freight containers and intermodal portable tank containers can be involved in an accident that may result in a fire or hazardous materials release. This may involve solids, liquids, or gases. In the case of intermodal freight containers, the scenario may involve numerous types of packages, containers, or cylinders.

The critical elements in analyzing a hazmat incident involving an intermodal container are (1) understanding the relationship between the hazardous material involved and its container, and (2) the impact the incident has upon the surrounding environment.

In this section we will focus primarily on the IM-101, IM-102 (International—IMO Type 1 and 2–T Codes T-1–T-22) and intermodal freight containers since they are most frequently encountered by emergency responders. We will also review emergency response issues and concerns at marine terminals, fixed facilities, and for each respective mode of transportation.

## Intermodal Portable Tank Container Emergencies

Based upon incident experience, the integrity of intermodal portable tank containers in transportation accidents has been fairly good. Most leaks involving portable tank containers have involved loose flanges, valves, blown burst discs, etc., rather than a breach of the container shell.

### ■ Damage Assessment

Intermodal tank containers are tough and hold up well in high-impact crashes such as a truck rollover or train derailment. (See **FIGURE 8-2**.) There are some basic factors that must be evaluated in all incidents involving intermodal tank containers, regardless of their location or the mode of transportation involved. If the container is involved in a rollover or derailment, the risks associated with the incident can rise.

As part of the size-up process, emergency responders must evaluate the following factors:

- Type of intermodal tank involved. For example, the IM-101, IM-102 (International—IMO Type 1 and 2–T Codes T-1–T-22) tank containers or the Specification 51, IMO Type 5 (T Code T50) tank container.

**FIGURE 8-1** In severe weather, containers can fall over-board off the ship or be carried away by flood water.
Courtesy of Hildebrand and Noll Associates, Inc.

- Whether the container is pressurized or nonpressurized and the type of tank metal (e.g., aluminum, stainless steel).
- Nature of the emergency. Scenarios could include leaking attachments, derailment, rollover, struck by object, etc.
- Container stress applied to the container tank; (e.g., mechanical, chemical, thermal, or a combination). Specific evaluation factors can include stress to the container jacket (i.e., insulation), mechanical damage to the tank frame, and corrosion to the container and/or frame.
- Type and nature of tank damage. For example, puncture, leaking manhole, bottom valve failure. If the container hasn't opened up or breached, responders should review the likely types of

**FIGURE 8-2** Tank frames on intermodal tank containers are very strong. They usually protect the tank from damage.
Courtesy of Greg Noll.

container breach and where the hazardous material will go once it escapes from its container. This is an integral element of the size-up process. Again, the shipper will usually be the best source of technical information regarding the container and likely breach and release scenarios.

- Amount of product both released and remaining in the container. The maximum amount of product contained in an intermodal tank container is approximately 6,300 gallons. Responders must evaluate where the incident is now, and where the incident will be once tactical operations are implemented. Experience shows that the most common releases from tank containers involve the valves and fittings. When safely possible, these releases can often be controlled by tightening the valve flange or bolts.

Jacketed and insulated containers can pose certain issues. If the jacket is damaged and the damage is limited to the outer jacket, the strength of the container may not be compromised. If liquid product is found escaping from the jacket, the actual source of the leak on the container may be at another location which is remote from the visible point of release.

When dealing with Specification 51, IMO Type 5 (T Code T50) containers, container damage assessment is critical due to the high container pressures involved. Remember—the higher the internal pressure, the farther the container and product will travel when breached and released. Also, liquefied gases commonly transported in Spec. 51 containers have high liquid-to-vapor expansion ratios. Special attention must be given to dents with sharp edges or gouges that cross over the heat-effected zone of welds or remove the upper bead of the weld. In these situations, the shipper should be consulted for guidance and advice before moving the tank container.

### ■ IM-101, IM-102 (International—IMO Type 1 and 2—T Codes T-1–T-22) Leak Control

Like most other types of hazardous materials emergencies, the majority of the spill and leak situations encountered involve loading and unloading accidents. Common scenarios include loose fittings and valves, overfills caused by product expansion, and "mystery leaks" through the tank shell. (See **FIGURE 8-3**.)

- **Overfills**—Overfills can be caused when a tank container is overfilled with product without room for expansion. Once the container is subjected to ambient heating after sitting in the hot sun for a while, the product may begin to overflow through the manhole cover or pressure relief devices may

**FIGURE 8-3** The primary source of leaks on intermodal portable tank containers comes from the foot valve shown at the bottom of the tank and the manhole cover on top of the tank.

Courtesy of Hildebrand and Noll Associates, Inc.

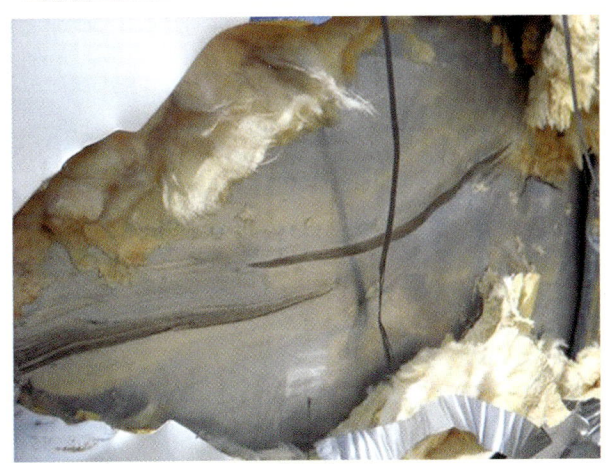

**FIGURE 8-4** This intermodal portable tank container has had its outer aluminum jacket ripped open. Note the layer of insulation and the insulation between the jacket and the tank shell. The tank shell has been damaged but not breached.

Courtesy of Hildebrand and Noll Associates, Inc.

actuate. In this situation, the container must be cooled to reduce the internal pressure and the effects of the ambient heating. As appropriate, product may have to be transferred from the container to reduce the potential for the problem to happen again, and burst discs may need to be replaced.

- **Mystery Leaks**—Another common scenario with tank containers are "mystery leaks" through the tank shell. For example, someone notices liquid dripping from an opening or crack through the insulating jacket in the lower half of the tank. If the tank has not been involved in an accident, the source of the leak is often water which has formed from condensation on the tank. If the outer jacket has been torn from a previous bump or scrape, the inner tank shell is exposed to warm air and sweats. The insulation between the inner tank and outer jacket becomes saturated with water and eventually drips out through openings in the outer jacket. If it is a warm day and the tank has an opening in the outer jacket, and the container has not been involved in an accident, you probably have water on the ground. Standard hazardous materials identification procedures, air monitoring, and sampling will usually take the mystery out of the situation. (See **FIGURE 8-4** .)
- **Bottom Outlet Leaks**—If product is leaking from the bottom outlet valve cap or a blind flange, make sure that the cap or flange is tight.

  Do not remove the valve cap without first checking the position of the valve. The valve

handle should be in the closed position—horizontal to the valve. If the valve handle is in the open position (in line with the valve), the remote emergency shutdown handle can be activated to close the internal foot valve. (See **FIGURE 8-5** .)

Intermodal tank containers are equipped with a remote shutoff device. The bottom outlet valve can be closed by activating the device. (See **FIGURE 8-6** .)

- **Container Shell Leaks**—If a tank container is damaged and the inner tank has been breached, the insulating jacket will make leak control tactics extremely difficult to implement. In this regard, jacketed tank containers share the same problems as jacketed MC-307/DOT-407 and MC-312/DOT-412 cargo tank trucks.

  If the tank container is a single-shell container, standard leak control tactics and techniques can usually slow the leak. Critical factors determining the success of patching and plugging operations will include the pressure of the leak (i.e., higher pressures = less effective leak control operations) and compatibility between the chemical and the patching/plugging device.
- **Top Fittings and Manhole Leaks**—In rollover situations the topside manhole cover may leak because the wing nuts were not properly tightened. These leaks are sometimes caused when the manhole dome cover wing nuts are tightened sequentially in a clockwise fashion rather than in an alternating fashion (e.g., the way you tighten lug bolts on an automobile wheel so the rim and tire

Flange bolted to tank

Butterfly valve in closed position

Valve cap

(A)

Courtesy of Glen Rudner.

(B)

Courtesy of Hildebrand and Noll Associates, Inc.

**FIGURE 8-5** (A) A butterfly foot valve attached to an intermodal portable tank container. The flange is bolted to the tank shell. The valve handle is across the valve and in the closed position. The threaded valve cap is in place. Use caution in removing the valve cap. There may be product trapped between the closed valve and the valve cap. (B) A foot valve removed from the container.

are straight). When this situation is encountered the leak can sometimes be stopped by alternately tightening down the wing nuts. (See **FIGURE 8-7** .)

## Intermodal Freight Container Emergencies

Some emergency responders refer to incidents involving intermodal freight containers as "surprise packages," because you're never quite sure what you will have until you open up the doors! (See **FIGURE 8-8** .)

From an emergency response perspective, freight containers present similar problems as vans and tractor-trailers. Remember the following safety practices:

- NEVER stand in front of freight container doors when opening the container. Unlock the door and using ropes or pike poles, stand off to the side as the container doors are opened. If anything goes wrong, responders will be off to the side and out of the high hazard area. (See **FIGURE 8-9** .)
- Loads may shift and come rolling out as the doors are opened. A 55-gallon drum of water weighs approximately 465 pounds, and most corrosives and other chemicals will weigh in excess of 500 pounds. Insulated containers can be very tight; opening the container doors and allowing the outside air to mix with the inside

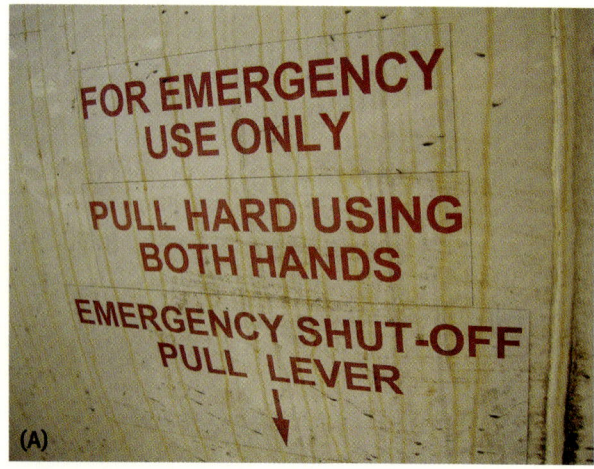

(A)

(B)

**FIGURE 8-6** (A) If a tank container is equipped with a remote shutoff it will be indicated with a marking on the tank shell. (B) To activate the remote shutoff, pull the cable.

Courtesy of Hildebrand and Noll Associates, Inc.

**FIGURE 8-7** The top fittings and manhole cover can be the source of a leak, especially if the container has rolled over. To stop a leak from the manhole cover, tighten the wing nuts clockwise.

Courtesy of Hildebrand and Noll Associates, Inc.

reactive environment may result in an explosion. (See **FIGURE 8-10**.)

- Consider checking for vent holes in freight containers, through which atmospheric monitoring may be performed with flexible tubing. Responders should also be aware of the use fumigants inside containers, such as methyl bromide, which are heavier than air and may remain in a container until a door is opened.
- If the freight container or trailer is completely loaded it may be extremely difficult to identify which container is leaking and is the source of the problem. In this case, responders will have to perform "container triage"—inspecting, removing,

**FIGURE 8-8** Intermodal freight containers contain both hazardous and nonhazardous cargo.

Courtesy of Hildebrand and Noll Associates, Inc.

and separating the containers one-by-one until the leaker can be identified.

- Forklift trucks or other container handling equipment may be required to perform the container triage process. If the container or trailer is intact, evaluate the hazards and risks with moving the vehicle to another location where it will be easier to manage the problem (e.g., off of the highway onto an adjoining road, isolated area of a truck stop, state highway maintenance area).

If a trailer has been involved in an overturn situation, forcible entry may be required. When dealing with noninsulated trailers, the easiest entry points will be through the (1) doors (sliding vs. swinging), (2) roof, (3) walls, and (4) floor. Remember that the walls may have metal or wood sheeting on the interior and that the floor is the strongest structural element of the trailer. (See **FIGURE 8-12**.)

The easiest access point will be through the doors. To open the container doors, saws or hydraulic tools may be required to cut the doorkeepers. Access through the walls or roof will require the use of saws with a carbide tip blade, while access through the floor will be extremely difficult. As with trailers, the floor is the strongest structural element of the freight container. (See **FIGURE 8-13**.)

Intermodal freight containers are used for transporting various types of regulated and nonregulated materials. Hazardous materials may be found in nonbulk packages such as drums, pails, bags, or cylinders, as well as in bulk containers such as IBCs or totes. If there is evidence of a leak, or the interior contents need to be inspected for damage, the box container needs to be opened. (See **FIGURE 8-14**.)

The best way to gain entry is to simply open the door using safe entry procedures. However, when box containers are involved in an accident or rollover, the container can be crushed or collapsed to the point that the container doors are jammed.

When the container cannot be accessed through normal means, forcible entry is required. Unlike the smooth-side aluminum or fiberglass containers, the corrugated container can be very difficult to open up if it is heavily damaged. The exterior walls are made from corrugated welded steel panels, and the interior structural members are well braced. In short, corrugated steel containers are tough!

One of the fastest ways to gain entry to a damaged corrugated steel container is to cut through the side or roof with a power saw using a carbide tip blade. However, saws can seldom be used due to the risk of fire or explosion from sparks.

Remember that even if the risk of explosion is not present, saws can throw sparks into the interior space through the saw cut and ignite combustible packing materials. As

**FIGURE 8-9** Intermodal freight container doors should always be secured before opening them regardless if it is for inspection or during a response.

Courtesy of Hildebrand and Noll Associates, Inc.

a general rule, saws are not a good forcible entry choice for steel containers because of the risk from fire. Saws may be appropriate for cutting if the freight container is on fire, where the doors can be opened and accessed for the fire attack. If the container cannot be opened via the doors, access holes can be cut into the sides or roof using a saw.

This method works when the container is loaded with Class A combustible materials (e.g., cardboard, cotton bales) and a smoldering fire needs to be extinguished. A triangular shaped hole can be cut and a hand line placed through the hole as a way to smother the fire before opening the container doors. (See **FIGURE 8-15** .)

**FIGURE 8-10** A loaded 55-gallon drum can weigh 450 pounds (204 kg). Use caution when opening freight container doors.

Courtesy of Hildebrand and Noll Associates, Inc.

**FIGURE 8-11** Each container may have to be removed before discovering the source of the leak.

Courtesy of Hildebrand and Noll Associates, Inc.

to minimize the risk of fire or explosion. Standard hazard and risk assessment and site safety procedures must be followed (e.g., monitoring, hazard control zones, protective clothing).

Hydraulic rescue tool spreaders equipped with the typical automotive tips are ineffective in forcing entry through the box doors. The wedge-shaped automotive tip is too large to spread the cargo doors open. When used to force the door's locking rod open, the wedge tip either pulled the entire rod assembly from the box or jammed the door further into the box. The tool can slip out creating a physical safety hazard. (See **SCAN SHEET 8-A**.)

## Industrial Facility Emergencies

Industrial facilities sometimes use intermodal containers to bring in specialty products such as additives, intermediates, and solvents. These are considered temporary containers, usually on site for less than 30 days. They may also be connected to process equipment.

Emergencies involving portable tank containers present emergency responders with the same general problems encountered at any facility where bulk storage tanks are involved. Nevertheless, there are some special problems that responders should be aware of when intermodal tanks are present. These include:

- **Exposures**—What type of exposures are there near the intermodal tank? Is the intermodal tank located in an isolated area or has it been incorporated into a process unit? Tanks connected to a chemical process unit through piping and valves may increase risks to responders. Look at the "big picture" to determine how the container and its contents may be affected and how it is being used.

**FIGURE 8-12** If forcible entry requires going through the roof, place two ladders on opposite sides so there are two ways off the roof. Entry through the roof should be considered a last resort. Regardless of what placards and markings indicate, you never know what you will find inside until you gain access.

Courtesy of Hildebrand and Noll Associates, Inc.

Hydraulic powered rescue tools are a good choice for forcible entry because they minimize the risk of fire. Remember, however, that the rescue tool's power plant is a potential ignition source when flammable liquids or gases are involved. In this case, the power unit must be placed as far away as possible from the container in order

Courtesy of Hildebrand and Noll Associates, Inc.

Courtesy of Glen Rudner.

**FIGURE 8-13** The fastest access into an intermodal freight container is through the doors. If the container doors have been damaged, forcible entry tools will be required. Remember to secure the doors with a chain or strap.

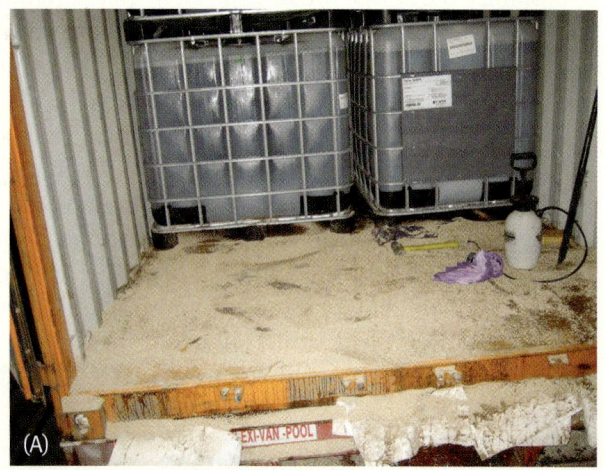

**FIGURE 8-14** (A) Intermediate bulk containers (IBCs) can contain 300 to 400 gallons (1,135 to 1,892 L) of liquid. Removing them requires a fork lift. An absorbent has been applied to the wooden flooring. (B) A 55-gallon drum has leaked paint. Note the catch basin.

Courtesy of Hildebrand and Noll Associates, Inc.

- **Hazard Class**—What type of hazardous material is in the container? OSHA regulations require that intermodal containers must continue to display their DOT placards onsite even though the container is not in transit. Remember that the IM 101 and IM 102 can carry a variety of different types of liquid hazardous materials (e.g., pesticides, flammable liquids, oxidizers). The Spec. 51 container is a pressure vessel with a slightly different container profile than the IM 102 and IM 102 containers. It will transport various products at higher pressures, including LPG and anhydrous ammonia. Use your basic responder skills—do not assume what is in

the tank. Recognize, identify, and verify before taking action.

- **Spill Containment**—What type of spill containment systems are in place? As previously noted, intermodal tanks are often used as temporary storage tanks. While fire codes require containment for permanent storage, spill control systems may not be in place. Anticipate a release and be aware of where personnel and vehicles are positioned. Be aware of the location of any drains, sewers, etc. into which the release may flow.

- **Code Enforcement**—If the intermodal tank container contains a flammable liquid, there are limitations on how long the container can be staged

**FIGURE 8-15** (A) Intermediate bulk containers can carry heavy fuel loads that can create fires that are difficult to extinguish and overhaul. (B) An initial fire attack should begin from the exterior. Hand lines should be ready before the doors are open. Beware of flashover as the doors open.

Courtesy of Hildebrand and Noll Associates, Inc.

on site. There are also restrictions on how close temporary containers can be to structures. See the Flammable and Combustible Liquids Code (NFPA 30) for guidance. Some local codes or ordinances may also require the use of the NFPA 704 marking system.

## Container Terminal Emergencies

Intermodal container terminals are very large facilities and can be confusing for emergency responders who have never toured or preplanned the facility. (See **FIGURE 8-16**.)

Modern terminal complexes can cover 600 to 1,500 acres and have thousands of containers parked, staged for loading, or in transit. A guided tour of any container terminal is a real education for emergency responders. You name it, and it is out there somewhere. (See **FIGURE 8-17**.)

Emergency response concerns at an intermodal terminal facility are a hybrid of the emergency response problems and concerns found within each mode of transportation. Specific issues and concerns include the following:

- Major ports and terminal areas are extremely busy, with a significant amount of traffic and container movement. Failure to pay attention to one's surroundings and daydreaming can easily get an emergency responder injured or killed. (See **FIGURE 8-18**.) For example, containers are constantly being loaded and unloaded, yardhorses are moving containers throughout the terminal, cranes and stackers are at work, and so forth. Remember to stay alert!
- Emergencies at port and terminal facilities can have significant economic and operational impacts

**FIGURE 8-17** The Barber's Cut container terminal in Houston, Texas receives and ships thousands of containers daily. A bar coding system helps identify where they are located.

Courtesy of Glen Rudner.

upon the facility. If an intermodal container is involved in an emergency, terminal personnel may move the involved container to a pre-identified, isolated location where the problem can be handled without disrupting other terminal operations. (See **FIGURE 8-19**.) As part of pre-incident planning activities, responders should identify these locations and evaluate site safety conditions, including surrounding exposures, ground contours, spill control, overall site safety, etc.

The container terminal Security and Safety Departments will be key players in the management of any emergency situation at seaports and rail terminals.

At marine terminals, the U.S. Coast Guard Captain of the Port (COTP) or a designated representative will coordinate all USCG operations. (See **FIGURE 8-20**.) Personnel from the USCG Marine Safety Office (MSO) will be key players in any hazmat emergency. The COTP has the authority to regulate and control the movement of both vessels and personnel within their area of responsibility, including denying vessels entry into port, prohibiting departure, placing specific operating requirements on vessels, and establishing restricted areas. The authority of the COTP also extends over the land-side areas of all waterfront facilities, such as terminals, piers, and wharves.

Locating and identifying containers holding hazardous materials at marine terminals can be difficult. While some shipments may not be placarded under DOT regulations, international shipments can be marked and placarded in accordance with IMDG regulations from the point of origin to the final destination, provided the trip is by an ocean carrier.

**FIGURE 8-16** If a container terminal is in your first due area, you should preplan the facility and know access points and be familiar with safe operating procedures.

Courtesy of Hildebrand and Noll Associates, Inc.

**FIGURE 8-18**   Container terminals are a beehive of activity when container ships are being loaded and unloaded. There is constant traffic. You must be aware of your surroundings at all times to stay safe.

Courtesy of Hildebrand and Noll Associates, Inc.

## Highway Transportation Emergencies

The highway mode of transportation can present a number of the emergency response issues. These include the following:

- Emergency responders have encountered numerous problems in dealing with both intermodal freight containers and tractor-trailers containing both hazardous and nonhazardous materials, which do not require placarding if the shipment is under 1,000 pounds of hazardous materials. Shipping documents may refer to these shipments as "freight of all kinds" (FAK), and may not accurately reflect what is inside of the container.

- Hazardous materials shipments originating outside of the United States can sometimes create problems in clearly identifying (1) who the shipper is, and (2) who the emergency response point-of-notification is within the United States. Although this situation has improved with the DOT Emergency Response Communication regulations, responders still can encounter problems in quickly accessing a knowledgeable individual who can provide accurate information on what is inside a freight container. In some instances, the U.S. point-of-contact has been a shipping broker who has little or no knowledge of the container contents. If in doubt, start with CHEMTREC® or other emergency response numbers that are required to be listed on shipping documents.

**FIGURE 8-19**   Container terminal operators may move "problem containers" to a pre-designated container isolation location within the terminal so that cargo operations can continue. Time is money when loading and unloading a container ship.

Courtesy of Hildebrand and Noll Associates, Inc.

**FIGURE 8-20**   The U.S. Coast Guard is a key player in any emergency that occurs at a marine terminal.

Courtesy of Hildebrand and Noll Associates, Inc.

**FIGURE 8-21** Hazardous materials shipments originating outside of the United States can create problems in clearly identifying exactly what is in the container.

Courtesy of Hildebrand and Noll Associates, Inc.

- While the majority of shippers and intermodal carrier companies are very reputable, there are also those who "live on the edge"—both operationally and financially. Hazmat enforcement and motor carrier inspections have consistently found problems with the mechanical integrity of equipment, as well as inadequate vehicle and trailer maintenance (e.g., brakes, connectors)
- Hazmat emergencies involving intermodal freight containers sometimes occur due to improper loading procedures. These can range from poor blocking and bracing to improper load separations between reactive chemical families and hazard classes. It is not uncommon to find the last row of containers "wedged" or pushed by a forklift into the freight container so that there is sufficient room to close the container doors. Normal movement and vibration during transportation can create mechanical stress upon the containers, eventually resulting in container leaks and spills. Murphy's Law dictates that the container, which is leaking, will be at the front and on the bottom of the freight container!
- Intermodal carriers, particularly those operating in port and terminal areas, will typically make multiple runs during a one- to four-week period. Several public safety Hazardous Materials Response Teams have experienced incidents where the driver had copies of the shipping papers in his cab for all runs during the preceding month. If the driver is seriously injured or hospitalized, it will take an extended period of time for emergency responders to gather, review, and determine which are the proper shipping documents.

- As previously noted, the ADR/RID Marking System assists European emergency responders in identifying the contents of a hazardous materials transportation container or vehicle. Since intermodal tanks and containers shipped into the United States and North America may contain these markings, emergency responders should have a basic understanding of this system. Information on the ADR/RID Marking System is listed in the *DOT Emergency Response Guidebook*. See Chapter 2 for more information on the ADR/RID Marking System.

## Railroad Transportation Emergencies

Remember the basic clues and their priority for identifying intermodal containers. (See **FIGURE 8-22**.)

- Container shapes.
- Markings and colors.
- Shipping papers and related documents. The train consist will be in the possession of the train crew.
- Description of each hazardous material being transported within the trailer or flat car.
- Senses (observation, noise, smells).

### ■ Trailer On A Flat Car and Container On A Flat Car

When dealing with TOFCs and COFCs, the consist will provide the following (See **FIGURE 8-23**):

- Number of the flat cars.
- Number of each trailer or container being transported on the flat car.

**FIGURE 8-22** Arriving at the scene of an incident, and using basic HazMat identification clues, you should be able to quickly size up that: 1) This is an intermodal tank container on its side with little damage and no visible leaks. 2) There are no placards or dangerous cargo markings. 3) It is probably an IM 101 container carrying nonhazardous cargo. You could develop this information without getting close to the container.

Courtesy of Hildebrand and Noll Associates, Inc.

**FIGURE 8-23** (A) Trailer On A Flat Car (TOFC). (B) Container On A Flat Car.
Courtesy of Hildebrand and Noll Associates, Inc.

- If the rail car is a TOFC, the shipping paper may be several pages long, with the required shipping paper entries for each hazardous material.

The latest generation of deep-well flat cars allows up to four 20-foot containers to be double-stacked. While these flat cars allow for the movement of more containers, they also create potential problems for emergency responders, including the following:

- If the COFCs on the bottom tier are placarded, the placards may be covered by the flat car framing and not be readily visible to emergency responders.
- Upper tier containers may be loaded "back to back" and be approximately 25 feet high, while the doors on the lower tier containers will be inaccessible due to the flat car's deep-well framing.

Flat cars containing COFCs may be connected in a series (sometimes referred to as "five packs"). To support both refrigerated or heated containers on these flat cars, there will be one central generator which supplies electricity. The generator is normally fueled by approximately 500 gallons of No. 1 or No. 2 diesel fuel oil. Electrical power is then supplied to the flat cars by 220V or 440V lines running within the flatcars.

Hazmat releases in TOFCs may pose a number of challenges for emergency responders, including the following:

- Trailers with swinging doors and which are loaded "back to back" will be difficult to open without removing the trailer from the flat car. Even if the trailer has a rear rolling door, limited space will make access extremely difficult.
- Even if the doors are accessible from the rear of the flat car, the door will be approximately 7 feet above grade. A ground ladder will then be required to unlock the doors. NEVER stand in front of the doors when opening the container.

Once the doors are unlocked, ropes or pike poles should be used while standing off to the side as the container doors are opened. If anything goes wrong, responders will be off to the side and out of the high hazard area.

- If the TOFC is involved in a derailment or rollover situation, there may be a failure of the fifth wheel attachment onto the flat car. Even if the attachment does not fail, due to the lightweight construction of modern trailers, there will likely be trailer roof and/or sidewall failure.
- If forcible entry into a TOFC is required, the easiest access points will be through the (1) doors, (2) roof, (3) walls, and (4) floor. Remember that the walls may have metal or wood sheeting on the interior and that the floor is the strongest structural element of the trailer.

## Marine Transportation Emergencies

Emergencies involving intermodal containers may occur while in port or underway. If the ship is in port and tied up at the container terminal, the ship's Master (Captain) and Chief Engineer are usually invaluable in understanding the vessel's design and equipment. They are also a reliable source of information concerning the ship's stability. All tactical plans involving operations on board ship must be fully coordinated with the Captain of the vessel. In U.S. ports, at least one member of the bridge management team is required to be able to converse in English. (See **FIGURE 8-24** .)

The Captain or First Officer will have a number of documents available to help identify the dangerous cargo location and contents. The ship's Officers will also be valuable in sizing up an incident and establishing the tactical action plan.

FIGURE 8-24 The bridge on a container ship is the "command post" during a shipboard emergency. In addition to communications, the Master or Mate can provide information in the Cargo Plan, Dangerous Cargo Manifest, and the Station Bill.
Courtesy of Hildebrand and Noll Associates, Inc.

FIGURE 8-25 Dangerous cargo emergencies at sea can be catastrophic with few options for the ship's crew.
© Alert Disaster Control (ALERT).

There are several useful documents on board the container ship that are available for emergency responders. These include:

- **Cargo Plan**— Lists the location of all cargo aboard the vessel.
- **Dangerous Cargo Manifest**—Required for all hazardous materials onboard. Usually found with the Captain or the First Officer on the bridge.
- **Station Bill**—Lists the duties and positions of every crewmember during an emergency. Written by the Captain, it is required under federal law to be posted in conspicuous locations throughout the vessel.

If an emergency occurs while in transit, the Captain basically has four options:

1. Handle the problem underway using the ship's crew or a land-based Emergency Response Team which joins the vessel en route.
2. Continue on to the original destination, which is usually a major port capable of handling the emergency using shore-side services.
3. Continue to the nearest available port or marina to either tie up or anchor offshore.
4. Abandon ship! Always an option of last resort and a real career ender for the Captain. (See FIGURE 8-25.)

The greatest potential for handling an emergency involving a shipboard intermodal container exists at a major port facility; however, intermodal containers can also be found on barges and smaller vessels in and around bays, rivers, canals, and lakes. (See FIGURE 8-26.) Many small coastal and inland ports and marinas are capable of receiving a small container ship or barge. If a fire or major hazmat emergency occurs while underway, the ship's Captain is primarily interested in assuring the safety of the crew and the vessel. If the vessel cannot safely reach its port of destination, the Captain may be forced to select an alternate destination much the same as an aircraft landing at the first available airport when an in-flight emergency occurs.

When a vessel declares an emergency at sea, the U.S. Coast Guard (USCG) will work with the ship's Captain to determine the best destination for assistance. Once notified, local response agencies usually have sufficient time to anticipate the problem and gear up for the ship's arrival.

FIGURE 8-26 Intermodal containers may be found anywhere on the waterfront including bays, rivers, and lakes.
Courtesy of Hildebrand and Noll Associates, Inc.

Bad experiences and surprises upon a vessel's arrival in port have prompted some ports and emergency response agencies to organize shipboard emergency response assessment teams. The typical Shipboard Assessment Team consists of five to ten specialists who are transported to the vessel by the U.S. Coast Guard while it is underway. Team members normally include a command officer, hazmat technician, paramedic, marine architect, USCG Marine Safety Office representative, and a marine engineer. The primary purpose of the team is to (1) size up the problem, and (2) determine what type of resources will be required to manage the incident when the vessel arrives at its destination. (See **FIGURE 8-27**.)

Shipboard emergencies involving hazardous materials and intermodal containers present many special problems to responders, including the following:

- **Loading Configuration**—Hazardous materials may be found above or below deck on the vessel. (See **FIGURE 8-28**.) Containers may range from portable tank containers to box containers loaded with a variety of different nonbulk packages and containers. When stacked in their transit configuration, the box containers placarded as hazmat containers can be difficult to distinguish from any other box container. Intermodal portable tank containers with hazardous cargo are usually stored on deck, near the stern of the ship aft of the super structure.

    Very few "working" hazmat incidents that involve intermodal containers can be resolved on board the ship. The nature of container ships is that box containers are stacked from the ship's keel to deck level

**FIGURE 8-28** Hazardous materials may be found above or below deck in an intermodal freight container on a container ship. Intermodal tank containers are usually loaded on the deck, aft of the super structure on the stern.

Courtesy of Glen Rudner.

and higher. Unless you are lucky enough to have the container on the top or side of the stack, there will be very few options to mitigate the problem. Basically, the ship must proceed to a major port, offload its cargo, and gain access to the problem container.

- **Confined Spaces**—The ship's cargo hold is a confined space. Accessing the problem may require climbing many levels below deck, with limited or no visibility, while the ship is pitching and rolling. All-in-all, a very dangerous situation.
- **Refrigerated Containers**—Refrigerated intermodal freight containers (also known in the marine industry as "snow boxes") are used for the transportation of foodstuffs and other temperature-sensitive items. They are packed with dry ice and generate carbon dioxide gas as the ice evaporates. Snow boxes can create special problems if they start to leak in confined areas (e.g., ship's hold) and create an oxygen-deficient atmosphere. Container companies recommend that "snow boxes" be stowed above deck only.
- **Language Difficulties**—The cost of increased regulation and taxes have forced many U.S. companies to register their vessels under a foreign flag. That means that a large number of vessels handling intermodal containers are sailing with foreign crews on board. While there are exceptions, the typical foreign crew does not speak English (at least not well enough to communicate during an emergency). Beyond basic shipboard firefighting, they have limited or no hazmat emergency response training, and they are primarily interested in getting off the ship when there is an emergency.

**FIGURE 8-27** A Maritime Emergency Response Team (MERT) can perform hazard and risk assessments on board the container ship before it is allowed to enter the port.

Courtesy of Hildebrand and Noll Associates, Inc.

# Scan 8-A—Forcing Entry into an Intermodal Freight Container

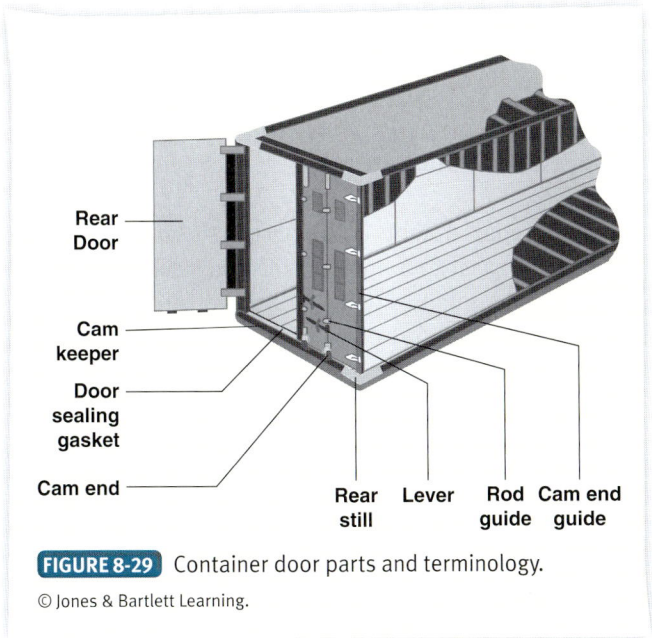

**FIGURE 8-29** Container door parts and terminology.
© Jones & Bartlett Learning.

Rear Door

Cam keeper

Door sealing gasket

Cam end

Rear still   Lever   Rod guide   Cam end guide

**FIGURE 8-30** Open the door latch by swinging the safety latch out of the way. Rotate the handle on the locking rod and see if the door swings open. If the door will not open, go to step 3.
Courtesy of Hildebrand and Noll Associates, Inc.

The fastest and most effective method of forcing entry is to use the power cutter to cut through the lock rod in four places.

The procedure is as follows:

1. As a safety precaution, a length of chain should be used to secure the doors in case containers are resting against the doors inside of the container. The load could fall onto the forcible entry crew. If a chain cannot secure the doors, they should be braced or shored up until forcible entry operations are completed and all personnel are clear of the rear doors.
2. Open the door latch by swinging the safety latch out of the way. Once the latch handle is free, try rotating the handle to free the lock rod. If the lock rod is jammed, use a pry bar to free the latch and rotate it out of its cradle. (See **FIGURE 8-30**.)
3. After freeing the lock rod latch from its cradle, use the hydraulic power cutter to cut through the lock rod approximately 8–10 inches from the top and bottom of the door. (See **FIGURE 8-31**.)

The lock rod is hollow steel construction and can be easily cut by a hydraulic tool. The top and bottom locking mechanisms are solid steel

**FIGURE 8-31** Cut the locking rods at 8–10 inches from the top and bottom of the door as indicated.
Courtesy of Hildebrand and Noll Associates, Inc.

construction and are too strong for the power cutter. Attempting to cut through the locking mechanism will be unsuccessful and could damage the cutter.

4. For best results, cut the right door lock rods first. The door may open without having to cut the left side lock rods. If the left-side door lock rods are cut first, you still have to cut the right-side door lock rods in both places, since the right door overlaps the left side door.

## Summary

Intermodal containers can be found at fixed facilities and in all modes of transportation. Consequently, both intermodal freight containers and intermodal portable tank containers can be involved in an accident that may result in a fire or hazardous materials release. This may involve solids, liquids, or gases. In the case of intermodal freight containers, the scenario may involve numerous types of packages, containers, or cylinders.

Intermodal Tank Containers are tough and hold up well in high impact crashes such as a truck rollover or train derailment. Like most other types of hazardous materials emergencies, the majority of the spill and leak situations encountered involve loading and unloading accidents.

5. Once all four lock rods have been cut cleanly through, a large pipe wrench should be used to grab hold of the door locking mechanism. Rotate the mechanism 90° so that the crow's foot is freed from its latch. This must be done at each of the four contact points. If the locking mechanism is not jammed, a Halligan Bar can be used, although a pipe wrench works well. REMEMBER—DO NOT COMPLETE THIS STEP UNTIL THE DOOR HAS BEEN BRACED OR SECURED.

Most leaks involving intermodal portable tank containers have involved loose flanges, valves, blown burst discs, etc., rather than a breach of the container shell.

Intermodal freight containers can involve almost any type of hazardous material. Dangerous cargo stored inside of the container will be packaged in Intermediate bulk containers (IBCs) or in small containers, packages, or cylinders.

## Reference

1. Noll, Gregory G. and Michael S. Hildebrand. 2014. *Hazardous Materials: Managing the Incident.* 4th ed. Burlington, MA: Jones and Bartlett Learning. pp. 155–156.

# Uprighting, Product Removal, and Transfer Operations

## CHAPTER 9

## Chapter Outline

- Key Terms
- Introduction
- Uprighting, Removal, and Transfer Operations
- Safety Measures
- Grounding and Bonding Considerations
- Product Transfer Methods
- Uprighting Methods
- Summary
- References

## Key Terms

**Bonding** A method of controlling ignition hazards from static electricity. It is the process of connecting two or more conductive objects together by means of a conductor.

**Grounding** A method of controlling ignition hazards from static electricity. The process of connecting one or more conductive objects to the ground through an earthing electrode (i.e., grounding rod).

**Risk-Based Response** A systematic process by which responders analyze a problem involving hazardous materials, assess the hazards, evaluate the potential consequences, and determine appropriate response actions based upon facts, science, and the circumstances of the incident (NFPA 472).

**Stabilization** Ensuring that all fires have been extinguished and ignition sources controlled, all spills have been confined, and the entire spill area has been foamed down, when required.

## Introduction

There are various types of intermodal containers that transport hazardous liquids, solids, gases, compressed gases, and refrigerated liquids. The most common type of container most emergency responders will encounter during an incident are nonpressure intermodal tank containers. Most Hazardous Materials Response Teams will have the capability to offload and transfer hazardous liquids from a damaged container with the assistance of an intermodal product or container specialist. This chapter discusses the damage assessment and safety factors that need to be considered before a nonpressure intermodal tank container is offloaded or uprighted.

Product removal operations cannot commence until after the incident site is stabilized. Stabilization means that all fires have been extinguished and ignition sources controlled, all spills have been confined, and the entire spill area has been foamed down, when necessary. The incident scene should be continuously monitored for flammability and toxicity.

## Uprighting, Removal, and Transfer Operations

Portable tank containers present the same challenges as cargo tank trucks and railroad tank cars when involved in an emergency. These can include: 1) how to assess the integrity of the container, 2) how to safely upright the portable tank container, and 3) how to safely offload the product. Transfer operations for upright containers will be easier than those for overturned containers. Guidelines for container uprighting, product removal, and transfer operations will be similar for all modes of transportation. (See **FIGURE 9-1**.)

### ■ Surveying the Container

The container must always be surveyed to determine: 1) the integrity of the container, 2) the safest method of uprighting, and 3) the safest method of offloading. This is particularly true when dealing with tank containers

**FIGURE 9-1** Portable tank containers can be involved in accidents that require offloading product in highway and rail accidents or at a marine terminal or fixed facility.

Courtesy of Hildebrand and Noll Associates, Inc.

that are loaded on a rail car and involved in a derailment or rollover situation.

There are no absolutes regarding whether to offload a container first or upright it while loaded. Each incident must be evaluated on its own merits using a risk-based response process. Factors to evaluate include the following:

- The type of tank container (nonpressure, pressurized, refrigerated).
- Nature of container stress. Damage assessment is critical. (See **FIGURE 9-2**.) If responders are unsure of the container damage or how a tank container is likely to breach, they should seek technical assistance from an intermodal product or container specialist. Source expertise may include the

**FIGURE 9-2** Containers should be inspected on all sides to look for damage. Damage can be superficial and only involve the outer tank shell.

Courtesy of California Governor's Office of Emergency Services.

shipper, carrier, railroad personnel, and tank container company representatives.

- The pitch and position of the container—front to back and left to right. It is possible that the container may move as product is pumped off and the product load shifts. Even when the container appears stable, consideration must be given to bracing. Bracing materials may include timber, jacks, or air bags.
- The position and location of the attachments that will be used for product offloading.
- The product being offloaded, including its properties and hazards (e.g., flammable liquid vs. fuming corrosive vs. poisonous gas).
- The level of training, resources, and equipment available for product transfer operations.

## Site Safety Measures

Removal operations cannot be initiated until after the incident site is stabilized. Stabilization criteria are that all spills and leaks have been controlled, any fires have been extinguished, and all ignition sources are controlled as necessary.

Specific site safety considerations which should be addressed during this phase of the incident include the following:

- Ensure that back-up crews with a minimum of 1-1/2-inch (38 mm) foam hand lines and at least two 20- to 30-pound (9.07 to 13.6 kg) dry chemical fire extinguishers are in place to protect all personnel involved in the offloading and uprighting operation. Some organizations will use a minimum of two foam hand lines, placing one foam hand line on each side of the unit for maximum protection. (See **FIGURE 9-3**.)
- Always have an escape signal and path for personnel working in the immediate hazard area.
- Air monitoring should be maintained throughout the duration of the incident. Action levels should be established and communicated to all on-scene personnel.
- Ensure that all personnel remain alert. Frequent rotation of personnel can usually address this need.

## Grounding and Bonding Considerations

When flammable liquids are involved, the generation and accumulation of static electrical charges during product transfer operations must always be considered. The foundation and justification for electrical grounding and bonding when flammable liquids are involved is

**FIGURE 9-3** When flammable or combustible liquids are involved, the spill area should be foamed before confinement operations begin. Air monitoring operations should continue throughout the duration of the incident.

© Jones & Bartlett Learning. Photographed by Glen E. Ellman.

well established in NFPA 472 – *Standard for Competence of Responders to Hazardous Materials/Weapons of Mass Destruction Incidents* (2013). The cargo tank must be grounded and bonded before product removal and transfer operations begin. **SCAN SHEET 9-A** provides more details on the grounding and bonding process.

General operational guidelines for bonding and grounding include the following:

- The area should be monitored to determine the concentration of flammable vapors. If the concentration is at or near the lower explosive limit (LEL), corrective actions will be required prior to transfer operations commencing.
- The pump and all pump-off appliances, such as hose couplings, downspouts, and recovery pans and "stingers," should all be bonded by connecting a bonding cable from the cargo tank to the appliance. In all appliance bonding operations, the first connection must always start at the damaged unit.
- Bonding cables must be placed on a clean, grease-free, paint-free surface. Cables with C-clamps are preferable to cables with "alligator clips" because they make better connections. At least five 50-foot

sections of 1/8-inch stainless steel grounding cables are suggested.

- Rubber hoses with a built-in wire will not necessarily provide bonding protection, as the wire within the hose may become broken or the wire may not be properly tied into the coupling.
- Plastic buckets can pick up static charges and should not be allowed for use as retention basins in an emergency situation.
- Grounding cables should initially be connected to the damaged container, then moved outward away from the overturned vehicles. The final connection can be made to a guard rail post or telephone or electrical pole support rod, providing it's deep enough to carry away the charge. Fire hydrants should not be used due to coatings on the ductile iron pipe.
- Grounding rod options include auger-type, T-handle grounding rods or 4–6 foot copper grounding rods. To increase the conductivity for grounding, some response teams pour water and salt around the grounding rod.
- An intrinsically safe ohm meter should be used to check the resistance at all connections. Periodically monitor all bonding and grounding cable connections to ensure that they remain in place and connected. To enhance operational safety, some response teams will sometimes provide double connections from the damaged cargo tank to ground.

## Product Transfer Methods

Product transfer and removal will normally be performed by product or container specialists working on behalf of the shipper or carrier, or environmental contractors. However, public safety responders will often continue to be responsible for site safety operations and will oversee the implementation of all transfer and removal operations.

There are three primary methods of product transfer for intermodal tank containers. These include:

1. **Portable Pumps**—Transfer operations between two intermodal portable tank containers using portable pumps may be either "open system" or "closed system." (See **FIGURE 9-4**.) Transfer pumps are categorized by their energy source and include gasoline, diesel, power take-off (PTO), electrical, water, and air-driven pumps. Key issues include:
   - Energy source and sparking potential of the pump. Gasoline, diesel, PTO, and electrical pumps can act as an ignition source in a flammable environment. Consider the hazardous

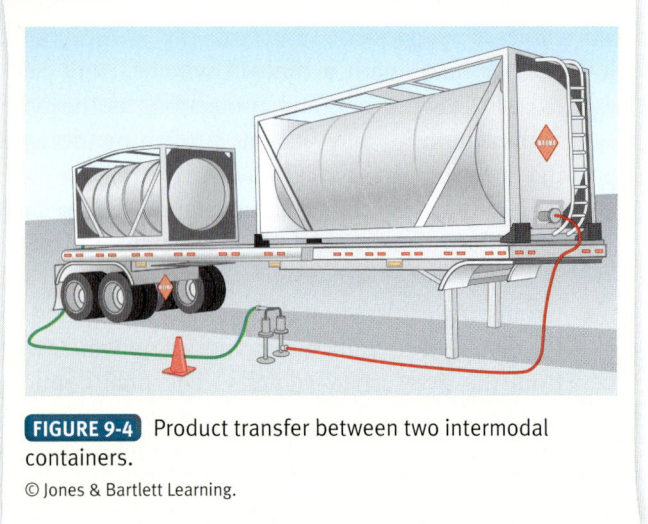

**FIGURE 9-4** Product transfer between two intermodal containers.
© Jones & Bartlett Learning.

classification ratings of any electrical equipment used around flammable liquids or gases. Class I, Division 2 should be the minimum accepted rating.

- Hazards of the material being transferred (e.g., flammable, corrosive, oxidizer).
- Materials of construction. The chemical compatibility of the pump, receiving tank, and all associated hoses and piping are critical factors. Responders should determine previous products handled by the equipment to ensure that there are no residual chemical contamination or reactivity hazards. In addition, product contamination may also be a concern, particularly when dealing with industrial solvents and other high quality chemicals.
- Power rating and pressure capacity of the pump including lift and flow capacities.

2. **Pressure Differential**—Vapors and gases may be transferred through the use of vapor compressors and pressure differential. Materials will seek the path of least resistance and naturally flow from high-pressure to low-pressure areas. In some cases, a compressor may be used in conjunction with a liquid pump to accelerate the rate of transfer by withdrawing vapors from the receiving container, compressing them, and forcing them into the damaged container. However, use of the vapor compressor will cause an increase in pressure within the damaged container and should only occur when the pressure increase can be safely accepted.

Inert gases, such as nitrogen and carbon dioxide, which are compatible with the product can also be used to move the contents of a damaged container into a receiving container. The inert pressure creates a positive pressure differential in the damaged container that pushes the liquid product into the receiving tank. Vapors from the receiving tank may have to be vented to the atmosphere or scrubbed.

3. **Vacuum Trucks**—These are frequently used to remove liquid hazardous materials and hazardous waste from an emergency scene. Depending on their rating and design features, they can handle flammable and combustible liquids, corrosives, and some poisons. (See **FIGURE 9-5**.)

A sewer or septic tank pumper is not a vacuum truck rated for hazardous waste. Vacuum trucks are designed for specific operating pressures up to 25 pounds per square inch gage (1.7 bar). A vacuum truck generally loads and unloads by reversing its vacuum pump through a four-way valve and manifold, which provides vacuum for loading and pressure for unloading. Some trucks are also equipped with a gear rotary pump for transfer operations. Vacuum trucks may be constructed to MC-312 / DOT-412 specifications.

Vacuum trucks must work close to the damaged hazmat container in order to reach pump-out connections or containment areas. Internal explosions within vacuum truck cargo tanks are rare. However, incidents that have occurred are usually due to pumping incompatible materials inside the tank (e.g., pumping a corrosive into a tank which is not rated for corrosives, or pumping

**FIGURE 9-5** Vacuum trucks may be used to remove hazardous liquids from intermodal tank containers.
Courtesy of Hildebrand and Noll Associates, Inc.

two incompatible hazardous chemicals into the same tank).

Vacuum trucks can be an ignition source and must be operated at the emergency scene with special precautions. Most vacuum truck fires and explosions are due to either operating the vehicle too close to the spill, pick-up, or discharge point, or failing to vent the vacuum pump discharge to a hazard-free area.

The following safety precautions should be observed when operating vacuum trucks within Hazard Control Zones:

- **Flammable Atmosphere Test**—Vacuum trucks should not be permitted to enter the hazardous area until flammable vapor monitoring has been conducted. While a reading of zero is preferred, it may not be realistic when open air spills are being recovered. As a general guideline, concentrations above 10% of the LEL are considered hazardous and may require additional response actions before transfer operations commence. Applying foam to a flammable liquid spill before beginning vacuum operations can help maintain the atmosphere within acceptable limits. Monitoring should also continue during the course of the transfer operation.
- **Grounding**—API Publication 2219—*Safe Operation of Vacuum Trucks in Petroleum Service* indicates that static electricity does not present an ignition problem with either conductive or nonconductive vacuum ruck hoses. However, with nonconductive hoses any exposed metal, such as a hose flange, can accumulate static electricity and act as an ignition source if the metal touches or comes close to the ground. Since it is often difficult to distinguish between nonconductive and conductive hoses in the field, API recommends that all exposed metal be grounded when any hose is used in other than a closed system with tight connections at both ends of the hose.
- **Venting**—When flammable or toxic liquids are loaded into a vacuum truck, the vacuum pump exhaust should be vented downwind of the truck

by attaching a length of hose sufficient to reach an area that is free from hazards and personnel.

- **Personnel Safety**—All unnecessary personnel should leave the area during loading. The vacuum truck driver should leave the truck cab. Strict control of ignition sources should be maintained within 100 feet of the truck, the discharge of the vacuum pump, or any other vapor source. Appropriate personal protective equipment should be used.

## Uprighting Methods

There is no best strategy to handle the uprighting of intermodal tank containers. A risk-based response process should be used in evaluating alternative options, and safety should be the number one factor in making the decision for the best course of action. The decision to offload or upright an intermodal tank container involving any type of hazmat requires careful consideration and input from a variety of technical specialists. These should include both product and cargo tank specialists and may also include rigging and heavy equipment specialists. All of these groups should be consulted before a plan of action is implemented.

If the intermodal tank container is being transported by truck, and the vehicle rolls over, the tank container may remain attached to the trailer or it may be detached.

Tank containers should not be lifted by equipment contacting the shell or any part of the frame other than the corner casings or corner posts. If there are any questions regarding the structural integrity of the container, responders should consult the shipper or tank container owner for additional guidance and recommendations.

Do not use the forks of a conventional forklift to lift a container. In addition, do not lift tank containers with equipment that contacts the shell or any part of the frame other than the corner castings or corner posts.

Lift loaded tank containers only after consultation with carrier and tank owners. Do not lift empty containers by the side or end frames *unless* the lower portion of the frame is still attached to the bottom corner castings.

# Scan 9-A—Grounding and Bonding Sequence

*What is bonding?* Bonding is the process of connecting two or more conductive objects together by means of a conductor, such as using an approved bonding wire to connect an aircraft being refueled to the fuel truck. Bonding is done to minimize potential differences between conductive objects, thereby minimizing or eliminating the chance of static sparking.

*What is grounding?* Grounding is the process of connecting one or more conductive objects to the ground through an earthing electrode (i.e., grounding rod). For example, connecting an aircraft to the ground through an approved grounding wire and connecting the fuel truck to the ground through a separate grounding wire and grounding rod. Grounding is done to minimize potential differences between objects and the ground. An ohm meter is used to measure the electrical resistance and ensure the electrical continuity of bonding and grounding operations.

*What is static electricity?* Static electricity is an accumulated electrical charge. In order for static electricity to act as an ignition source, four conditions must be fulfilled:

1. *There must be an effective means of static generation.* This can occur when a flammable or combustible liquid is moved from one place to another through pipes, filtering, or by pouring. Some products like gasoline are good static accumulators and can pick up a static charge as they pass through piping during loading operations. Products that easily accumulate static charges must be loaded at slower flow rates to permit downstream relaxation time for the product to lose its charge.
2. *There must be a means of accumulating the static charge buildup.* Not every product lends itself to accumulating a static charge.

3. *There must be a spark discharge of adequate energy to serve as an ignition source (i.e., incendive spark).* We have all experienced a static discharge at one time or another when we walked across a carpet during the winter and touched a metal object or exited a car in winter when the humidity is low, wearing a nylon jacket or wool slacks, and then touched the car door. The spark can be seen jumping between your finger and the car's metal. Not every static spark carries enough energy to cause ignition, and even if it does, there must be an adequate flammable mixture in air present (see item 4 below).
4. *The spark must occur in a flammable mixture.* In order for a fire or explosion to occur in the presence of an adequate ignition source (static spark), the fuel-to-air mixture must fall within the flammable range. By bonding and grounding, you are giving a static charge a pathway in which to travel to earth without creating a spark. The resistance of the grounding field will be affected by weather, type of soil, moisture content of the soil, and the time of year.

## ■ Electrical Resistance Guidance

Emergency response organizations should adopt an acceptable electrical resistance level in their Standard Operating Procedures for grounding purposes. For example, the National Electrical Code (NEC) notes that the ground level should be <25 ohms resistance for "residential purposes," a standard that has been adopted by many emergency response agencies.

In 2014, the NFPA 77 Committee on Recommended Practice on Static Electricity changed the recommended resistance to 1,000 ohms. Section 7.4.1.3.1.1 of NFPA 77 states: *"In field-based situations such as hazardous materials response operations or flammable/combustible* materials spill control and transfer, it might be necessary to establish a temporary or emergency grounding system in a remote location in order to dissipate static charges. In such situations, various types of conductive grounding rods, plates, and wires, are sometimes used in combination to increase surface area contact with the earth. If the purpose of the temporary grounding system is to dissipate static electricity, a total resistance of up to 1 kohm (1,000 ohms) in the ground path to earth is considered adequate. This can be measured using standard ground resistance testing instruments and is realistically and quickly achievable in most types of terrain and weather conditions." There are some variables, such as incident location and soil type, that may make achieving 1,000 ohms difficult.

## ■ Grounding and Bonding Sequence for an Overturned Intermodal Portable Tank Container

**FIGURE 9-6** shows the grounding and bonding sequence for an overturned cargo tank truck. The cargo tank truck tractor may be connected to the cargo tank trailer via the fifth wheel or it may have separated from the cargo tank. The cargo tank metal is in contact with the ground. In this configuration, the product must be removed before the cargo tank can be uprighted. This illustration does not depict the proximity or exact spatial layout of the bonding and grounding system, rather, it shows the sequence of the various connections.

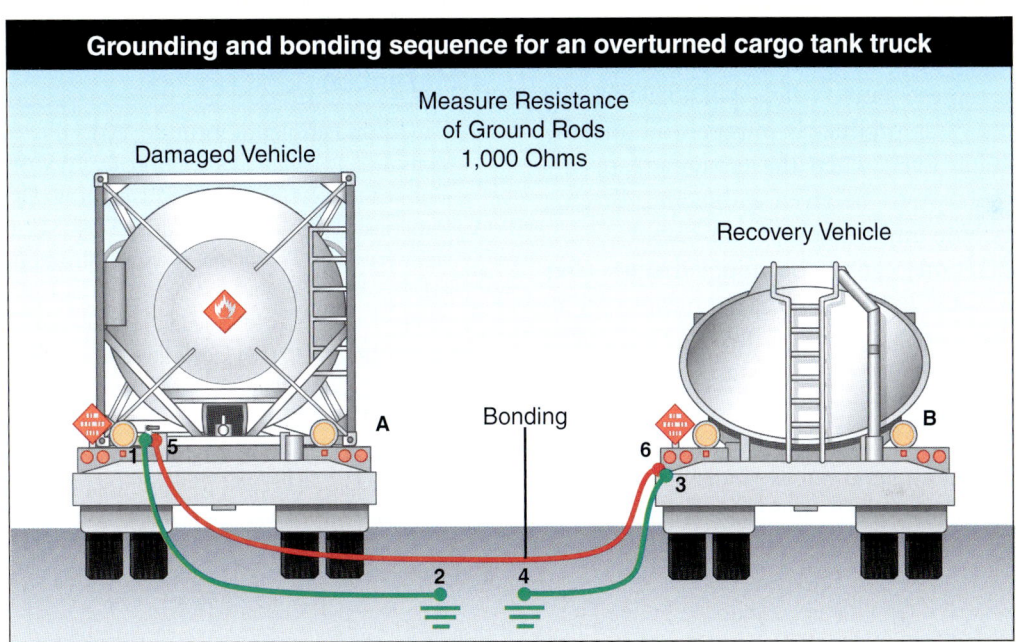

**FIGURE 9-6** Intermodal tank containers are equipped with earth connections for bonding and grounding. The locations are marked. These connections are typically rusty and need to be wire brushed to clean metal to ensure a good connection.

© Jones & Bartlett Learning

## Summary

Intermodal portable tank containers present the same challenges as cargo tank containers and railroad tank cars when involved in an emergency. These can include: 1) how to assess the integrity of the container, 2) how to safely upright the container, and 3) how to safely offload the product.

Before offloading or uprighting operations begin, the container must always be surveyed to determine: 1) the integrity of the container, 2) the safest method of uprighting, and 3) the safest method of offloading. A risk-based response process should be used for this survey and evaluation process. This is particularly true when dealing with tank containers involved in a derailment or rollover situation.

Damage assessment is critical. If responders are unsure of the container damage or how a tank container is likely to breach, get technical assistance. Sources may include the shipper, railroad personnel, and tank container company representatives. It is possible that the container may move as product is pumped off and the product load shifts. Even when the container appears stable, consideration must be given to bracing. Bracing materials may include timber, jacks, or air bags.

Removal operations cannot be initiated until after the incident site is stabilized. Stabilization means that all spills and leaks have been controlled, any fires have been extinguished, and all ignition sources are controlled as necessary.

The generation and accumulation of static electrical charges during flammable liquid transfer operations must always be considered. To minimize the potential of a flash fire or explosion, this static build-up must be controlled through bonding and grounding.

Product transfer and removal will normally be performed by the shipper or environmental contractors, but public safety responders will often continue to be responsible for site safety operations and will oversee the implementation of all transfer and removal operations.

There are three primary methods of product transfer for intermodal tank containers. These include: 1) portable pumps, 2) pressure differential, 3) vacuum trucks.

A risk-based response process should be used in evaluating alternative options, and safety should be the number one factor in making the decision for the best course of action. The decision to offload or upright an intermodal tank container involving any type of hazmat requires careful consideration and input from a variety of technical specialists.

## References

1. American Petroleum Institute. 2015. *Protection Against Ignitions Arising Out of Static, Lightening, and Stray Currents.* API Publication 2003. 8th ed. Washington, DC.
2. American Petroleum Institute. 2005. *Safe Operations of Vacuum Trucks in Petroleum Service.* API Publication 2219. 3rd ed. Washington, DC.
3. Noll, Gregory G., Michael S. Hildebrand, Glen Rudner, and Rob Schnepp. 2014. *Hazardous Materials: Managing the Incident.* 4th ed. Burlington, MA: Jones and Bartlett Learning. pp. 378–384.
4. Union Pacific Railroad Company. *A General Guide to Tank Containers, Participant Manual.* (September 2006). Omaha, Nebraska. Appendix A, pp. 4–19.

# Appendix

## Acronyms

**AAR** Association of American Railroads

**ACEP** Approved Continuous Examination Program

**ADR** European Agreement Concerning Transport of Dangerous Goods by Truck

**API** American Petroleum Institute

**ASME** American Society of Mechanical Engineers

**ATA** American Trucking Association

**BEA** Bureau of Economic Analysis

**BIC** Bureau International des Containers et du Transport International

**BLEVE** Boiling Liquid Expanding Vapor Explosion

**BSP** British Standard Pipe

**BSW** British Standard Whitworth

**BTS** Bureau of Transportation Statistics

**CBP** U.S. Customs and Border Protection

**CBRNE** Chemical, Biological, Radiological, Nuclear, Explosive

**CHEMTREC** Chemical Transportation Emergency Center

**COC** Convention for Safe Containers

**COFC** Container On A Flat Car

**COPT** Captain of the Port

**DOT** Department of Transportation

**DWT** Deadweight Ton

**EMC** Emergency Action Code

**FEU** Forty-Foot Equivalent Container Unit

**FHWA** Federal Highway Administration

**FRA** Federal Railroad Administration

**HTS** Harmonized Tariff Schedule

**ICC** International Chamber of Commerce

**IED** Improvised Explosive Device

**IMDG** International Maritime Dangerous Goods Code

**ISO** International Standards Organization

**ITCO** International Tank Container Organisation

**JOC** Journal of Commerce

**MARAD** Maritime Administration

**MEG** Multiple Element Gas Container

**NCHRP** National Cooperative Highway Research Program

**NFPA** National Fire Protection Association

**NOM** Normas Oficiales Mexicanas

**NRC** National Research Council

**OSHA** Occupational Safety and Health Administration

**PIERS** Port Import and Export Reporting Service

**PTO** Power Take-Off

**RID** European Agreement Concerning Transport of Dangerous Goods by Rail

**SOLAS** Safety of Life at Sea

**TC** Transport Canada

**TEU** Twenty-Foot Equivalent Container Unit

**TOFC** Trailer On A Flat Car

**TRB** Transportation Research Board

**TSA** Transportation Security Administration

**TSI** Transportation Services Index

**TTI** Texas Transportation Institute

**TWIC** Transportation Worker Identification Credential

**UN** United Nations

**UNECE** United Nations Economic Commission for Europe

**USCG** United States Coast Guard

**USDHS** U.S. Department of Homeland Security

**USDOC** U.S. Department of Commerce

**USDOE** U.S. Department of Energy

**USDOT** U.S. Department of Transportation

**WMD** Weapons of Mass Destruction

# Glossary

**Bill of Lading** A document that establishes the terms of a contract between a shipper and a transportation company. It serves as a document of title, a contract of carriage and a receipt for goods.

**Blocking** Wood or metal supports to keep shipments in place to prevent cargo shifting. See also bracing.

**Bonding** A method of controlling ignition hazards from static electricity. It is the process of connecting two or more conductive objects together by means of a conductor.

**Box** A container with structural framework and panel members fastened together to form a rigid enclosure. The panels used to create this enclosure can be made of corrugated paper, plywood, or any product strong enough to perform containment of given products. Most boxes are fully enclosed and can have any section (i.e., side, end, top, base, and cap) removable for filling.

**Bracing** Wood or metal supports to keep shipments in place to prevent cargo shifting. See also blocking.

**Break-Bulk Cargo** Packages of maritime cargo that are handled individually, palletized, or unitized for purposes of transportation as opposed to bulk and containerized freight.

**Bulk Cargo** Refers to freight, both dry or liquid, that is not packaged, such as minerals (oil, coal, iron ore) and grains. It often requires the use of specialized ships such as oil tankers as well as specialized transhipment and storage facilities. Conventionally, this cargo has a single origin, destination, and client. It is also prone to economies of scale.

**Cargo** Freight loaded into a ship.

**Cargo Manifest** A manifest that lists all cargo carried on a specific vessel voyage.

**Carrier** Any person, organization, or government undertaking the transport of dangerous goods by any means of transport. The term includes both carriers for hire or reward (known as common or contract carriers in some countries) and carriers on own account (known as private carriers in some countries).

**Chassis** A frame with wheels and container-locking devices in order to secure the container for movement.

**Cladding** The outer skin around the shell of an intermodal tank container that protects the insulation from water. Usually made from GRP (glass reinforced plastic) or aluminum. Also known as a tank jacket in the United States.

**Commodity** The article shipped. For dangerous and hazardous cargo, the correct commodity identification is critical.

**Container** A large, standard-size metal box into which cargo is packed for shipment aboard specially configured oceangoing containerships. It is designed to be moved with common handling equipment to enable high-speed intermodal transfers in economically large units between ships, railcars, truck chassis, and barges using a minimum of labor. Therefore, the container rather than the cargo in it serves as the transfer unit.

**Container On A Flat Car (COFC)** An intermodal container shipped on a railroad flat car.

**Containerization** A system of intermodal freight transportation that uses standard containers that can be loaded onto vessels, railcars, and trucks. It involves the stowage of general or special cargo in a container for transport in the various modes.

**Container Markings** Markings on both freight containers and portable tank containers that provide useful information for safe container and cargo handling, as well as potential safety or environmental hazards.

**Container Port** A harbor with marine terminal facilities for transferring cargo between container ships and land transportation, such as truck or rail.

**Container Ship** A cargo vessel designed and constructed to transport, within specifically designed cells, portable tanks and freight containers which are lifted on and off the vessel with their contents intact. There are two types of container ships: full and partial. Full container ships are equipped with permanent container cells with little or no space for other types of cargo.

Partial container ships are considered multipurpose container vessels, where one or more but not all compartments are fitted with permanent container cells, and the remaining compartments are used for other types of cargo. This category also includes container/car carriers, container/rail car carriers, and container/roll-on/roll-off vessels.

Crate  A container with structural framework fastened together to form a rigid structure enclosure. Typically they have an open construction concept with little or no panel support.

Cryogenic Tank Containers  Containers that transport refrigerated liquefied gases such as argon, helium, oxygen, nitrogen, and ethylene. They include the UN Portable Tank T75 and IMO Type 7 specification containers.

Cylinder  A transportable pressure receptacle of a water capacity not exceeding 39.63 gallons (150 L).

Damage Assessment  The process of gathering and evaluating container damage as a result of a hazardous materials incident.

Dangerous Goods  As defined by the United Nations and IMDG Code, dangerous goods are materials or items with hazardous properties which, if not properly controlled, present a potential hazard to human health and safety, infrastructure, and/ or their means of transport.

Data Plate  A metal identification plate that shows many details of the tank container including its specification, current test data, the manufacturer, and the permitted weight/capacity.

Document Tube  A sealed tube fitted to the frame of the intermodal portable tank container for carriage of pertinent documents like Certificate of Analysis (COA) and a cleaning certificate. Also known as a document holder.

Draft  The number of feet that the hull of a ship is beneath the surface of the water.

Drum  A flat-ended or convex-ended cylindrical packaging made of metal, fiberboard, plastic, plywood, or other suitable materials. This definition also includes packagings of other shapes (e.g., round taper-necked packagings, or pail-shaped packagings). Wooden barrels or jerricans are not covered by this definition.

Dry Bulk Cargo  Cargo that may be loose, granular, free-flowing, or solid, and is shipped in bulk rather than in package form. Dry bulk cargo is usually handled by specialized mechanical handling equipment at specially designed dry bulk terminals.

Dunnage  Packaging materials used to keep cargo in place inside a container or transportation vehicle.

Emergency Response  Response to any occurrence which has or could potentially result in the release of a hazardous substance.

Flat Car  A freight car that has a floor without any housing or body above. Frequently used to carry containers and/or trailers or oversized/odd-shaped commodities. The three types of flat cars used in intermodal are conventional, spine, and stack cars.

Flexible Intermediate Bulk Container  A flexible sack or bag manufactured from woven fabric, cloth, or a plastic material with a capacity of as much as 2,000 lbs.

Frames  International Standards Organization (ISO) dimensioned frames on intermodal portable tank containers that are typically fabricated from high tensile carbon steel and comply with ISO length, width, and height dimensions.

Freight Container  An article of transport equipment that is of a permanent character and accordingly strong enough to be suitable for repeated use; specially designed to facilitate the transport of goods, by one or other modes of transport, without intermediate reloading; designed to be secured and/or readily handled; having fittings for these purposes; and approved in accordance with the International Convention for Safe Containers (CSC), 1972, as amended. The term "freight container" includes neither vehicle nor packaging. However, a freight container that is carried on a chassis is included. For freight containers for the transport of Class 7 material, a freight container may be used as a packaging. A small freight container is that which has either any overall outer dimension less than 4.9 feet (1.5 m), or an internal volume of not more than 3 cubic meters. Any other freight container is considered to be a large freight container.

Fusible Device  A non-reclosing pressure relief device that is thermally activated and provides protection against excessive pressure buildup in the portable tank developed by exposure to heat, such as from a fire.

General Cargo  General cargo consists of those products or commodities—such as timber, structural steel, rolled newsprint, concrete forms, and agricultural equipment—that are not conducive to packaging or unitization. Break-bulk cargo (e.g., packaged products such as lubricants and cereal) are often regarded as a subdivision of general cargo.

Gross Weight  The total weight of a container plus the maximum weight of the cargo that can be carried inside.

Grounding  A method of controlling ignition hazards from static electricity. The process of connecting one or more conductive objects to the ground through an earthing electrode (i.e., grounding rod).

Hazardous Materials  As defined by the U.S. DOT [49 CFR 171.8], any substance or material capable of posing an unreasonable risk to health, safety, and property when transported in commerce. This includes hazardous substances, hazardous wastes, marine pollutants, and elevated temperature materials.

**Intermediate Bulk Container (IBC)** Any rigid or flexible portable packaging that has a capacity of:

1. Not more than 3.0 cubic meters, 793 gallons (3,000 L) for solids and liquids of packing groups II and III;
2. Not more than 1.5 cubic meters for solids of packing group I when packed in flexible or rigid plastics, composite, fiberboard, and wooden IBCs;
3. Not more than 3.0 cubic meters for solids of packing group I when packed in metal IBCs;
4. Not more than 3.0 cubic meters for radioactive material of Class 7;
5. Is designed for mechanical handling;
6. Is resistant to the stresses produced in handling and transport, as determined by tests.

**Intermodal** Used to denote movements of cargo containers interchangeably between transport modes (i.e., motor, water, and air carriers) and where the equipment is compatible within the multiple systems.

**Intermodal Container** A standardized shipping container built to transport materials in multiple modes – road, rail, water, and air – without unloading and reloading the cargo, fitted with a rigid frame with corner castings for tie-down and lifting to facilitate mechanical handling with its contents intact.

**Intermodal Portable Tank Container** An intermodal container for transporting liquids, solids, and gases in bulk with a capacity of 118.9 gallons (450 L) or more. The tank is capable of being loaded and discharged without the need for removal of its structural equipment. It has stabilizing members external to the shell, and is capable of being lifted when full. It is designed primarily to be loaded on to a vehicle or vessel and is equipped with skids, mountings, or accessories to facilitate mechanical handling.

**Jacket** The outer insulation cover or cladding around the outside of an intermodal tank container which may be part of the insulation.

**Leak Control** Actions taken to contain or keep a material, which could pose a threat to health, safety, and/or the environment, within its container.

**Manhole** Located on the top of the intermodal portable tank container. It is used to enter the tank for internal inspection or top loading.

**Marine Terminal** A designated area of a port, which includes but is not limited to wharves, warehouses, covered and open storage spaces, cold storage plants, grain elevators and bulk cargo loading and unloading structures, landings, and receiving stations, used for the transmission, care, and convenience of cargo and/or passengers in the interchange of same between land and water carriers or between two water carriers.

**Multiple Element Gas Container (MEGC)** Assemblies of UN cylinders, tubes, or bundles of cylinders interconnected by a manifold and assembled within a rigid frame with corner castings for the transport of gases. Also known as a tube module.

**Net Weight** The weight of the goods that are to be shipped, not including the packaging. The gross weight is the total weight, including the goods and the packaging. These numbers are often needed to calculate fees and taxes.

**Nonpressure Intermodal Tank Containers** Standardized 20-foot (6.058 m) bulk liquid stainless steel vessels supported and protected by a steel frame that can easily be moved from ship to highway or rail transport vehicles and can be stacked for storage or transit.

**Outer Packaging** The outer protection of a composite or combination packaging together with any absorbent materials, cushioning, and any other components necessary to contain and protect inner receptacles or inner packaging.

**Packaging** One or more receptacles and any other components or materials necessary for the receptacles to perform their containment and other safety functions.

**Placards** A display that conveys information about the hazards of the bulk cargo on board a motor vehicle, rail car, or intermodal container. This includes the hazard class and sometimes the four-digit identification number that allows emergency responders to identify the material and obtain emergency initial response guidance and information from the DOT Emergency Response Guidebook. Placards are required to be displayed on both sides and at each end of the motor vehicle or rail car.

**Port** A harbor area in which marine terminal facilities for transferring cargo between ships and land transportation are located.

**Portable Tank** (a) For the purposes of the transport of substances of Class 1 and Classes 3 to 9, a multimodal portable tank. It includes a shell fitted with service equipment and structural equipment necessary for the transport of dangerous substances; (b) For the purposes of transport of nonrefrigerated, liquefied gases of Class 2, a multimodal tank having a capacity of more than 118.9 gallons (450 L). It includes a shell fitted with service equipment and structural equipment necessary for the transport of gases; (c) For the purposes of transport of refrigerated liquefied gases, a thermally insulated tank having a capacity of more than 118.9 gallons (450 L) fitted with service equipment and structural equipment necessary for the transport of refrigerated liquefied gases. The portable tank shall be capable of being loaded and discharged without the need of removal of its structural equipment. It shall possess stabilizing members external to the shell, and shall be capable of being lifted when full. It shall be designed primarily to be loaded onto a vehicle or vessel and is equipped with skids, mountings, or accessories to facilitate mechanical handling. Road cargo tank-vehicles, rail cars nonmetallic tanks, gas cylinders, large receptacles, and intermediate bulk containers (IBCs) are not considered to fall within this definition

Pressure Relief Valve  A pressure (or sometimes a combination pressure/vacuum) relief device that is fitted to every intermodal portable tank container tank to protect the tank against excessive over pressure or vacuum. Also known as a vacuum relief valve.

Pressure Tank Containers  Containers designed to handle liquefied gas and other high vapor-pressure products. These pressurized tank containers include:

1. Tank containers for nonrefrigerated liquefied compressed gases (UN portable tank T50, DOT Specification 51, IMO Type 5)
2. Refrigerated liquefied gases (cryogenic tank containers—UN portable tank T75, DOT Specification 51, IMO Type 5)
3. Nonrefrigerated compressed gases (tube modules such as Multiple Energetic Gas Containers - MECG).

Risk-Based Response  A systematic process by which responders analyze a problem involving hazardous materials, assess the hazards, evaluate the potential consequences, and determine appropriate response actions based upon facts, science, and the circumstances of the incident (NFPA 472).

Shell  The part of the portable tank which retains the hazardous materials intended for transportation, including openings and closures, but does not include service equipment or external structural equipment.

Stabilization  The point in a hazardous materials incident at which the adverse behavior of the hazardous material is controlled. It involves ensuring that all fires have been extinguished and ignition sources controlled, all spills have been confined, and the entire spill area has been foamed down, when required.

Tank  A tank container, a road tank-vehicle, a rail tank-wagon, or a receptacle to contain solids, liquids, or gases, with a capacity of not less than 450 L when used for the transport of gases.

Tank Container  An intermodal container for transporting liquids, solids, and gases in bulk with a capacity of 118.9 gallons (450 L) or more and includes Multiple Element Gas Containers (MEGC), also called tube modules, which are the high-pressure equivalent of tank containers.

Tare Weight  The weight of an empty vehicle or container. By subtracting the empty weight from the gross weight (laden weight), the weight of the goods carried (the net weight) may be determined.

Trailer On A Flat Car  Truck trailers shipped on a railroad flat car.

Transportation Worker Identification Credential (TWIC)  A security program designed to ensure that individuals who pose a threat do not enter a marine terminal or a facility attached to a marine terminal or merchant marine vessel. The TWIC is issued by the U.S. Department of Homeland Security.

Vehicle  A road vehicle (including an articulated vehicle, i.e. a tractor and semi-trailer combination), railroad car, or railway wagon. Each trailer shall be considered as a separate vehicle.

Vessel  Any seagoing vessel or inland waterway craft used for carrying cargo.

Walkway  Ladder and top walkways on an intermodal portable tank container for an easy access to the top of the tank.

The following references were used to develop this glossary:

1. Noll, Gregory G. and Michael S. Hildebrand. 2014. *Hazardous Materials: Managing the Incident*. 4th ed. Burlington, MA: Jones and Bartlett Learning. pp. 460–486.
2. Rodrigue, John-Paul and Brian Slack. 2013. *The Geography of Transport Systems*. 3rd ed. New York: Routlege.
3. United Nations, 2015 *Recommendations on the Transport of Dangerous Goods - Model Regulations*. 19th ed. New York
4. U.S. Department of Transportation, Maritime Administration. 2008. "Glossary of Shipping Terms." Washington, DC.
5. National Fire Protection Association, NFPA 472: Standard for Competence of Responders to Hazardous Materials/Weapons of Mass Destruction, 2013, Quincy, MA.

# Index

© Photos.com/Getty

Note: The letter 'f' following locators refers to figures.

**A**

AAR 600 marking, 19–22
ADR/RID hazard marking system, 16–18
    container markings, 16–18
    emergency responses, 91
    digit (identification numbers) system, 26–29
    non-pressure intermodal tank containers, 51
air line connection, 56f
air monitoring operations, 99f
Association of American Railroads (AAR), 10–11

**B**

bags, 35–36
bar coding system, 89f
bonding, 98–99, 102–103, 103f
boxes, 2, 7, 35, 37–39, 39f, 40f, 72, 94
bracing materials, 98

**C**

calibration chart (strapping chart), 22
Chemical, Biological, Radiological, Nuclear, Explosive (CBRNE) prevention, 12
container markings
    AAR 600 marking, 19–22
    ADR/RID hazard marking system, 16–18
    calibration chart (strapping chart), 22
    check digit, 16f
    convention for safe containers (CSC), 21–22
    country, size, and type, 15
    customs convention for containers, 22
    data/identification plate, 13–15
    do not use fork truck decal, 24
    do not walk decal, 23–24
    European Railway System Approval Decal (UIC), 21
    food grade cargo decal, 24
    HAZCHEM marking system, 16
    height markings, 22
    IMDG Code Markings, 22
    ISO Standard 3166, 15
    ISO Standard 6346, 16
    owners code and number, 15–16, 18
    placards, 16
    for portable tank containers, 18–19
    TC impact marking, 21
    valve operation instructions, 24
Container On A Flat Car (COFCs)
    emergency responses, 91–92, 92f
    railroad operations, 70

container shell leaks, 83–84
container terminal emergencies, 89–90
Convention for Safe Containers (CSC), 8, 11, 21–22
cryogenic tank containers
    features, 65
    holding time, 64–65, 64f
customs convention for containers, 22

**D**

dangerous goods
    European Agreement Concerning Transport of Dangerous Goods by Rail (RID), 16–17
    European Agreement Concerning Transport of Dangerous Goods by Truck (ADR), 16
    hazard class, 16, 34
    International Maritime Dangerous Goods (IMDG) Code, 9, 51
    Transport of Dangerous Goods Directorate (TDG), 10
    UN's recommendations, 8–9, 10, 73
Department of Homeland Security (DHS), 8, 12
Department of Transportation (DOT)
    codes, standards and regulation, 8, 10, 11f, 13–15, 19, 21f
    definition, non-bulk packaging, 35
    emergency responses, 83, 88–91, 100
    gas containers, intermodal tank, 60–61, 63–64, 66–67
    non-pressure intermodal tank containers, inspection, 43–44, 48, 50–51, 57–58
    Pipeline and Hazardous Materials Safety Administration (PHMSA), 10
drums, 36–38, 37f, 85

**E**

Electrical Resistance Guidance, 102–103
emergency responses
    bottom outlet leaks, 83
    cargo emergencies, 93f
    container shell leaks, 83
    damage assessment, 81–82
    intermodal freight container, 84–87
    manhole leaks, 83–84
    mystery leaks, 83
    overfills, 82–83
    top fittings, 83–84
    useful documents, 93
European Agreement Concerning Transport of Dangerous Goods by Rail (RID), 16–17, 18f, 24f, 51f, 91

European Agreement Concerning Transport of Dangerous Goods by Truck (ADR), 16, 18f, 24f, 51f, 91
European Railway System Approval Decal (UIC), 21

**F**

flammable atmosphere test, 101
flexible intermediate bulk container (FIBC), 37–38, 41f
food grade cargo decal, 24f
freight container
    domestic transportation, 4
    PHMSA regulation, 10
    radiological screening, 12f
freight container, construction features
    bags and sacks, 35–36
    boxes, 37
    cargo capacities and weights, 34
    cargo packaging systems, 34–35
    container dimensions, 30–31
    container floors, 34
    corner castings, 34
    corrugated-side freight containers, 31–32
    drums, 36–38, 37f
    external-post containers, 33
    flexible intermediate bulk container (FIBC), 37–38
    nonbulk packaging, 35
    rigid intermediate bulk containers (IBC), 38–42
    smooth-side freight containers, 32–33

**G**

grounding, 98–99, 101, 102–103, 103f

**H**

HAZCHEM Marking System, 16
hazard class, 1, 5, 7, 13, 16, 18f, 34, 35, 43, 50, 57f, 59, 65, 88, 91
hazardous material
    AAR 600 intermodal portable model, 71
    cargo packaging system, 35
    cryogenic tank containers, 64
    definition, DOT, 8, 10
    drums, 37
    emergency responses, 55, 81–94
    flexible container, 38
    foot valves, 54
    intermediate bulk containers (IBCs), 40, 42, 96
    marking system, 13–29, 50
    Mexican regulation, 10–11
    tanks and valve testing, 58–59
    water reactive, 36

HazMat identification, 91f
highway transportation emergencies, 90–91

**I**

IMDG Code Markings, 22
IM 101 Portable Tanks (International—IMO Type 1)
    leak control, 82–84
    nonpressure intermodal tank, 51–52
IM 102 Portable Tanks (International—IMO Type 2)
    leak control, 82–84
    nonpressure intermodal tank, 52
improvised explosive device (IED), 12
industrial facility emergencies
    code enforcement, 88–89
    exposures, 87
    hazard class, 88
    spill containment, 88
Intelligence Reform and Prevention Act of 2006, 12
intermediate bulk containers (IBCs), 35, 37–38, 40–42, 88f
intermodal container
    handling requirements, 4–7
    international, internal, and industry standards, 9f
    land bridge concept, 4
    in modern ships, 5f
    modes of transportation, 3f
    operations, 68–70
    purpose, 2–4
    uses, 1–2
intermodal container ships
    cargo handling and stowage, 75
    loading methods, 77
    off-loading methods, 77–78
    permanent storage, 76–77
    safety guidelines for transfer operations, 78
    semipermanent storage, 76–77
    temporary storage, 75
    types, 73–75
intermodal freight container
    construction features, 30–42
    emergency responses, 84–87
    forcing entry, 95–96, 95f
    hazardous materials, 94
    target audience, 2
intermodal portable tank container, 2
    chassis configuration, 79f
    transportation, 71–72
International Maritime Dangerous Goods (IMDG) Code, 8–10, 12, 17, 22, 51, 56, 58, 89
International Maritime Organization (IMO), 5–6, 19, 43–44, 51–52, 59, 61
International Standards Organization (ISO), 5, 8, 11, 15–16, 30–31, 36, 44, 53, 65
    ISO Standard 1161, 34
    ISO Standard 6346, 25

**L**

land bridge concept, 4
Land Transportation Standards Hazardous Materials Working Group (LTSS Group 5), 10
leak control, 81–83

**M**

manhole cover, 5, 6, 53, 54f, 82, 83f, 85f
marine operations

crane technology, 72–73
emergency responses, 92–95
information management systems, 73
international standards, adoption of, 73
marine terminals, 73
removal of trade barriers, 73
upgraded ports, 73
Maritime Emergency Response Team (MERT), 94f
Maritime Transportation Security Act of 2002, 12
Mexican Regulation for the Land Transport of Hazardous Materials and Wastes, 10
Multiple Element Gas Containers (MEGCs), 65–67
mystery leaks, 83

**N**

National Electrical Code (NEC), 102
NFPA 77 Committee on Recommended Practice on Static Electricity (2014), 102
nonhazardous material, 48–49, 52, 85f, 90, 91f
nonpressure intermodal tank containers
    access to tops, 52
    air line connection, 56
    design and construction features, 43–44
    dimensions and capacities, 44
    dipstick, 53–54
    discretionary fittings, 56
    electrical controls, 46
    fittings, 52
    foot valves, 54–55
    frame type, 46–47
    grounding connections, 57
    heating units, 45–46
    inspections, 58
    insulation, 46
    linings, 45
    manhole cover, 53–54
    markings, 47–51
    materials, 46
    refrigeration units, 45
    safety devices, 57
    spill box, 52
    steam line connection, 56
    tank shape, 45
    testing, 58
    top-loading valves, 54
nonpressure intermodal tank containers, types
    IM 101 Portable Tanks (International—IMO Type 1), 51–42
    IM 102 Portable Tanks (International—IMO Type 1), 52

**O**

Official Mexican Standards (Normas Oficiales Mexicanas or NOMs), 10
overfills, 82–83

**P**

Pipeline and Hazardous Materials Safety Administration (PHMSA), 10–11
placards, 16, 17f, 22, 29, 34, 48f, 50–51, 58, 73, 87–88, 91–92
portable tank containers
    AAR 600 requirements, 11
    data plate marking, 13–15
    DOT special permit, 19

electrical resistance guidance, 102–103
grounding and bonding considerations, 98–99, 102–103, 103f
minimum tank design pressure, 19
overhead hazard marking, 19
PHMSA regulation, 10
product transfer methods, 99–101
site safety measures, 98
specification markings, 19
surveying, 97–98
unique markings, 18–19
uprighting methods, 101
weight markings, 19
pressure tank containers
    data plate, 60–61
    features, 60
    fittings, 61–63
    safety devices, 63–64
    UN T50 instructions, 59–60
product transfer methods
    portable pumps, 99–100
    pressure differential, 100
    vacuum trucks, 100–101

**R**

railroad operations
    container on a flat car (COFCs), 70
    equipment, 70
    trailer on a flat car (TOFCs), 70–71
railroad transportation emergencies, 91
refrigeration units, 45
rigid intermediate bulk containers (IBC), 38–42
risk-based response, 97

**S**

safety guidelines, for transfer operations, 78–79
site safety measures, 98
stabilization, 97
steam line connection, 56f

**T**

target audience, 7
TC impact marking, 21
Trailer On A Flat Car (TOFCs)
    emergency responses, 91–92, 92f
    railroad operations, 70–71
Transport Canada (TC), 10
transportation security regulations, 11–12
Transportation Worker Identification Credential (TWIC®), 12

**U**

U.S. Hazardous Materials Regulations, 10
United Nations Economic Commission for Europe (UNECE), 9

**V**

vacuum trucks, 100f
venting, 101

**W**

weight markings, 19